LASER SAFETY HANDBOOK

LASER SAFETY HANDBOOK

Alex Mallow
Leon Chabot

VNR **VAN NOSTRAND REINHOLD COMPANY**
NEW YORK   CINCINNATI   ATLANTA   DALLAS   SAN FRANCISCO
LONDON   TORONTO   MELBOURNE

Van Nostrand Reinhold Company Regional Offices:
New York   Cincinnati   Atlanta   Dallas   San Francisco

Van Nostrand Reinhold Company International Offices:
London   Toronto   Melbourne

Copyright © 1978 by Litton Educational Publishing, Inc.

Library of Congress Catalog Card Number: 78-1484
ISBN: 0-442-25092-4

Manufactured in the United States of America

Published by Van Nostrand Reinhold Company
135 West 50th Street, New York, N.Y. 10020

Published simultaneously in Canada by Van Nostrand Reinhold Ltd.

15 14 13 12 11 10 9 8 7 6 5 4 3 2 1

Library of Congress Cataloging in Publication Data

Mallow, Alex.
   Laser safety handbook.

   Includes bibliographical references and index.
   1. Lasers—Safety measures.   I. Chabot,
Leon, joint author.   II. Title.
TA1677.M34   621.36'6'028   78-1484
ISBN 0-442-25092-4

Preface

The purpose of this handbook is to organize and compile under one cover information on the many and diverse considerations that affect laser safety. Until now, laser safety information needed to describe the topic adequately has been scattered throughout a large collection of articles, pamphlets and brochures.

The contents of this handbook are divided into areas as follows:

Chapters 1 through 5 introduce the laser beam hazard and associated laser hazards. Biological effects of laser radiation are addressed. Laser theory is treated in a simplified fashion, free of complicated mathematics and concepts, yet adequate for an understanding and appreciation of the material that follows. The same approach has been used in dealing with laser measurements—simplified emphasis on those measurement requirements and techniques of use in laser safety.

Chapter 6 comprehensively addresses protective standards covering intrabeam and extended source viewing considerations. Maximum permissible exposure standards for eye and skin exposure to laser beams are provided. Correction factors to these standards and special handling techniques for complex emission variations are also included, along with problems and solutions.

Chapters 7 and 8 cover laser beam hazard evaluation and classification techniques, and the approach to control of the beam hazard. Control of associated laser hazards is discussed in Chapter 9.

Federal and state public laws concerned with laser safety are treated in Chapter 10, from both a survey and an in-depth approach.

Laser safety programs, while functionally a part of the hazard-control aspect, are addressed separately in Chapter 11, and the discussion includes several approaches to training programs. As an offshoot, safety in the classroom is included in Chapter 12, with emphasis on safety aids used in classroom laser experiments and demonstrations. Certainly, a suitable medical surveillance program and appropriate laser protective eyewear are vital elements of a laser safety program. These two elements are covered in Chapters 13 and 14, respectively.

Chapter 15 examines various aspects of atmospheric effects on laser beams.

Appendix A provides an extensive definition and laser terminology summary, which is of interest from both a laser and a laser safety viewpoint.

Appendix B contains a copy of New York State Industrial Code Rule 50, "Lasers." It has been included as a ready reference for the interpretation of this document covered in Chapter 10, Public Laws.

Finally, Appendix C, adapted from the Bureau of Radiological Health's (BRH) "Guide For Submission of Information on Lasers and Products Containing Lasers Pursuant to CFR 1002.10 and 1002.12," details the approach required of the laser manufacturer in reporting laser product accessible radiation levels.

ACKNOWLEDGMENTS

The following tables and figures contained in the handbook are reproduced with permission from "American National Standard for the Safe Use of Lasers, Z 136.1-1976," copyright 1976, by the American National Standards Institute, copies of which may by purchased from the American National Standards Institute at 1430 Broadway, New York, New York 10018: Tables 5-1 and 5-2, 6-1 to 6-4; Figures 6-4 to 6-13. Appreciation is extended to the Grumman Aerospace Corporation for having provided the authors with the opportunity of interesting themselves in laser safety, as a part of their responsibility in support of the A-6E TRAM flight development program. Finally, heartfelt appreciation is extended to Marilyn Mallow, for numerous hours spent in typing and proofreading the manuscript.

Commack, New York   A. Mallow
                                                           L. Chabot

Contents

1

Introduction to Laser Safety and Hazards

Since the invention of the laser some 15 years ago, a vigorous industry has emerged. There are many scientific, engineering and industrial laser applications—in surgery, material processing and machining, nuclear fusion experiments, construction, scanners for deciphering coded package markings and others uses. Military applications are common, and the communications field is making increased use of lasers for exploiting their tremendous potential for carrying very large amounts of information.

WHY CONCERN?

Lasers can be hazardous owing to the great brightness of the beam. The principal concern is with eye damage, since that organ is capable of increasing laser light intensity many thousands of times by its focusing power. With the ever increasing laser intensities encountered, skin damage is of no small concern.

Aside from the beam hazard, operation of laser systems like any other technological or industrial process involves various potential associated hazards. Electrical shock is a principal associated hazard and has probably been responsible for more accidents than has the laser beam. Other associated laser hazards include airborne contaminants, ionizing radiation, optical radiation excluding the laser beam, noise and so on.

PROTECTIVE STANDARDS FOR EYE AND SKIN EXPOSURE

Recognizing that the laser beam is potentially hazardous, suitable control of that hazard necessitates establishment of protective standards that provide criteria for allowable exposure of the eye and skin to laser radiation. Such standards (both voluntary and government-decreed) have evolved, based on the best available information from experimental studies, and are sometimes referred to as Maximum Permissible Exposure or MPE levels.

MPE determinations can become complex, as is the case when a number of widely different laser wavelengths are emitted or when pulses are superimposed on a CW background.

EVALUATION AND CONTROL OF THE LASER HAZARD

It follows that control of the laser hazard is best preceded by a full evaluation of the laser hazard itself. Further, for a complete hazard evaluation it is necessary to consider the complete range of laser hazards (e.g., electrical, and so forth), not only the beam hazard. A suitably organized and administered laser safety program, with the necessary safety training, will minimize the risk considerably.

The laser beam hazard evaluation is governed by the following three key features:

- The potential of the laser beam to injure personnel which is addressed using a standardized laser classification of the relative beam hazard (e.g., Class I–IV lasers).
- The laser environment.
- The personnel present in the laser environment.

The latter two items do not lend themselves to a standardized approach, as does the first item, because of the many environments the laser can be used in and the many situations in which personnel can be exposed to the laser beam.

Control measures for the laser beam hazard are strongly influenced by the classification of the laser being used. Clearly, a Class IV high-powered laser will require more stringent control measures than lasers of lesser power (and therefore lower classification). A good measure of experience and judgment is needed to implement a balanced array of engineering controls, administrative procedures and other control measures in order to adequately protect personnel, but not unduly stifle progress. The latter eventuality is a distinct possibility.

SUMMARY

The chapters and appendices that follow elaborate considerably on the key laser safety elements addressed above. As already noted, determination of MPEs can become complex, as can be the case for laser classifications. Thus, where deemed particularly useful, a number of problems and solutions are provided to assist the reader in better understanding and applying safety criteria and rules.

2
Basics of Lasers

INTRODUCTION

The letters in the word LASER come from the phrase *L*ight *A*mplification by *S*timulated *E*mission of *R*adiation.

The light generated by a laser is quite different from ordinary light, possessing advantages that for the most part cannot be obtained using other light sources.

Light is produced by internal atomic actions, and a particular form of these internal actions generates laser light. Assume for simplicity that an atom contains a small dense nucleus and one or more electrons in motion about the nucleus.

ELECTRON ENERGY LEVELS

The electrons exist at distinct energy levels with respect to the nucleus. The electrons normally occupy the ground state, which is the lowest energy level. By absorbing energy, the electrons can change to a higher energy level. By emitting energy, the electrons change back to the vacant lower level. Figure 2-1 illustrates an electron transition to a lower level with the energy delta, ΔE, emitted as a photon of light.

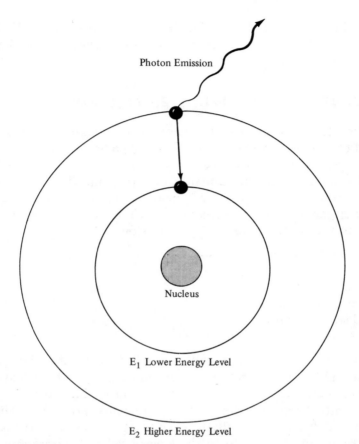

Photon Emission

Nucleus

E_1 Lower Energy Level

E_2 Higher Energy Level

Figure 2-1. Electron transition from a higher to a lower energy level, with resulting photon emission.

The upward transitions result when energy is absorbed from one of a possible number of sources. For laser action, two ways of supplying energy are quite important. The energy of a light photon can be transferred to an orbital electron. The electron will then move to a higher or "excited" level. The quantum nature of the atom requires that the electron take on only the exact amount of energy that will transfer it from one permitted energy level to another.

Electrical discharges are a second main technique used to excite electrons. Here the energy is supplied by collisions with electrons moving in an electric field. Whether photons or electric discharge techniques are used, an electron moves to a higher energy level as the atom is excited.

SPONTANEOUS AND STIMULATED EMISSION

The atom tends to revert to the lowest energy state. As a result, an excited electron in a higher energy level will go back to a lower energy level. Typically, an electron loses its excitation and reverts to a lower orbit by the spontaneous emission of a photon. The energy of the photon must equal the difference in energy between the excited and the lower energy level. The frequency of the released photon will be related to the energy delta by the expression.

$$E = hf$$

where

  E is the energy delta
  h is Planck's constant
  f is the frequency

The familiar formula $f = c/\lambda$ will be recalled, where c is the velocity of light and λ is the wavelength. Depending on type, lasers have output wavelengths between 0.2 micrometer and 1 millimeter.

The emitted photon released from an excited atom will, when interacting with another similarly excited atom, result in the second atom having an electron go to the lower level and release a photon. The process is called stimulated emission. The frequency, energy, direction and phase of the second photon will be the same as for the first photon; the first photon will continue on its original path, accompanied by the second photon. The two photons can now de-excite more atoms through this process called stimulated emission.

With spontaneous emission, the light output occurs more or less equally in all directions as the excited electrons fall randomly to the lower energy level. By comparison, stimulated emission causes an increase in the number of photons traveling in a particular direction. Indeed, a preferred direction can be established by placing highly reflective mirrors at the ends of an optical cavity. While photons not moving perpendicularly to the mirrors will escape, the number of photons traveling along the axis of the two mirrors increases owing to reflections back and forth and continuing stimulated emission. The number of photons increases greatly along the preferred path, i.e., large amplification occurs in the number of photons. To obtain efficient amplification, a condition of so-called population inversion is needed. More electrons must be excited to the upper energy levels than remain in the normal lower energy levels; then population inversion exists.

ELEMENTS OF A LASER

The three principal parts of a laser have been indicated in the preceding discussion. The first is a substance or medium in which to achieve the population inversion. The second is an input energy source (called the pump) to raise electrons to an excited state. The third is the low loss optical cavity used to establish the lasing axis and increase stimulated emission by means of repeated reflections.

Medium

A mixture of helium and neon is used as the medium in a common type of gas laser, and gives an intense red light. Another type of laser that uses gas as the medium is the CO_2 laser; the emitted radiation is in the far infrared and is not visible. The first successful laser was a solid state laser, which used a ruby rod with a controlled amount of chromium impurity as the lasing medium. Liquid organic dyes are used as lasers. A fourth type of laser is the semiconductor laser, consisting of a semiconductor junction. High efficiency and small size are characteristics of semiconductor lasers.

Energy Source (Pump)

Since lasing will occur efficiently when a population inversion has been established in the lasing medium, energy is pumped into the medium to bring about population inversion. Several typical methods of pumping exist.

Optical pumping (supplying light energy) is employed in solid state and liquid lasers. A powerful source of light is focused on the lasing medium. The photons supplied by the pumping that have the correct energy cause electrons of the lasing material to move to a higher level.[1] Xenon flashtubes are frequently used as optical pumps for solid state lasers. Liquid lasers are typically pumped by a beam from a solid state laser.

Gas lasers use electron collision pumping. An electrical discharge is sent through the gas-filled tube. The electrons traveling in the discharge current lose energy by colliding with gas atoms or molecules, and the atoms or molecules that receive energy are excited. A continuous laser output can result, since the collision pumping can be done continuously.

Optical Cavity

Laser action may begin if the lasing medium has been pumped and a population inversion obtained. However, photons will be produced in all directions if no control is placed over the direction of beam propagation. The use of an optical cavity can control the direction of beam propagation. A reflector is placed at each end of the lasing medium to form the cavity. Figure 2-2 is a schematic representation of a typical optical cavity.[2] The reflecting mirrors face each other and are aligned very carefully. Photons traveling along the cavity axis are reflected back and forth through the laser medium, causing more stimulated emission, which causes the beam to grow in strength. One of the mirrors is only partially reflecting and permits part of the beam to be transmitted out of the cavity. Gas lasers use mirrors as ends for the gas tube or can utilize external mirrors. The first ruby lasers were constructed with rod ends polished optically flat and silvered; semiconductor lasers use a similar technique.

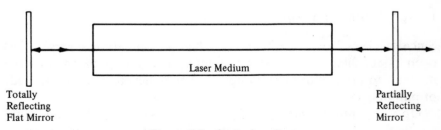

Totally
Reflecting
Flat Mirror

Laser Medium

Partially
Reflecting
Mirror

Figure 2-2. Optical cavity.

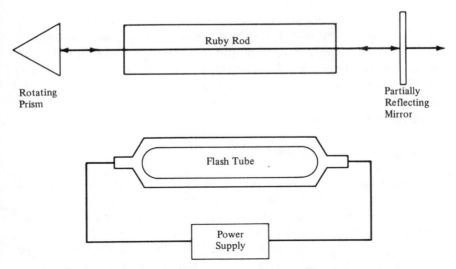

Figure 2-3. Schematic of solid state laser with optical pumping.

CONTINUOUS WAVE AND PULSE TYPE LASERS

With continuous pumping, equilibrium can be reached between the number of photons produced by atoms raised to the excited state and the number of photons emitted and lost. Continuous laser output results, but is generally used only with low power input levels. Higher power pumping usually is supplied in the form of a pulse, and the laser output is also in pulse form. Figure 2-3 shows a solid state laser with optical pumping.[2] The rotating prism causes an output pulse when it reaches an orientation that allows reflection back and forth in the cavity, and is an example of Q-switching, as compared to simple pulsing of a pump such as a flash lamp.

Q-SWITCHING

Q-Switching (or Q-spoiling) is employed to produce an exceptionally high-power output pulse. The optical cavity is actually a resonant device, and usage of the term "Q" stems from its use in electrical resonant circuits. Q-switching in lasers refers to the method by which the power of the laser emanates in short bursts. The Q-switch is a device that interrupts the optical cavity during pumping to prevent losses due to lasing which would otherwise start as soon as population inversion occurred and would last as long as pumping maintained the inversion. In this fashion the pumping achieves a higher level of population inversion. The Q switch is then made to restore the cavity to a resonant condition, and a very short and much more powerful pulse is achieved.

PROPERTIES OF LASERS

The energy of a laser is measured in joules (J), and laser power is measured in watts (W), where watts equal the number of joules per second. The output of lasers, particularly pulsed lasers, is frequently rated in joules/cm^2 or watts/cm^2 (e.g., J/cm^2 or W/cm^2).

The light energy emerging from a laser does not spread (diverge) very much at all. The energy is not greatly dispersed as the beam travels. Laser beam divergence is measured in milliradians. A typical laser can have a divergence of 0.5 to 1.0 milliradian. The width of a laser beam will be $(a + r\theta)$,[2,3]

where

    a = initial beam diameter, cm
    r = distance from laser, cm
    θ = beam divergence in milliradians, small θ

Two other important properties of lasers are monochromaticity and coherence. The term monochromatic means one color, or one wavelength of light. Laser light is very close to being monochromatic. The phrase spatial coherence is used to describe two or more waves of the same frequency, phase, amplitude and direction. Laser light comes very close to being perfectly spatially coherent.

REFERENCES

1. Van Pelt, W. F., Stewart, H. F., Peterson, R. W., Roberts, A. M., Worst, J. K., *Laser Fundamentals and Experiments*, U.S. Dept. of Health, Education, and Welfare, Public Health Service, Bureau of Radiological Health, Southwestern Radiological Health Laboratory, Las Vegas, Nevada, May 1970.
2. "Control of Hazards to Health from Laser Radiation," U.S. Department of the Army Technical Bulletin TB-MED-279, Washington, D.C., May 1975.

3. "American National Standard for the Safe Use of Lasers, Z136.1-1976," American National Standards Institute, New York, 1976.

BIBLIOGRAPHY

1. Duly, W. W., *CO$_2$ Lasers, Effects and Applications*, Academic Press, New York, 1976.
2. Marshall, S. L., Editor, *Laser Technology and Applications*, McGraw-Hill Book Company, New York, 1968.
3. "Laser Hazards and Safety in the Military Environment," AGARD Lecture Series No. 79, North Atlantic Treaty Organization, 1975.

3

Biological Effects of Laser Radiation

INTRODUCTION

Laser radiation striking human tissue directly, or reflected onto the tissue, is capable of inflicting varying degrees of damage. The principal concern is with eye damage, since ocular injuries can readily be of an irreversible nature. While the skin hazard is of a decidedly secondary consideration, it can be considerable.

This chapter includes descriptions of tissue damage mechanisms, as well as factors contributing to tissue injury. Although the content is generally applicable to both the eye and skin, the eye structures are mentioned with sufficient frequency to make minimal familiarity with this organ desirable. As such, the first topic covered is a brief description of the eye structure.[1,2]

Finally, the eye and skin hazards are addressed largely from the point of view of specific biologic effects. In the case of the eye, a summary tabular form relates specific biologic effects to particular eye structures and laser radiation spectral bands. For the skin, a brief description of this organ is followed by discussion of various biologic effects that can be caused by laser radiation.

DESCRIPTION OF THE EYE

A schematic diagram of the human eye is shown in Figure 3-1. The eyeball is spherically shaped, with most of its interior volume containing a clear viscous substance called the vitreous humor. Surrounding the vitreous humor are three basic layers:

- Outermost layer or sclera
- Middle layer comprising the choroid, ciliary muscle and the iris
- Inner layer or retina

The sclera is a tough white tissue that functions as the supporting framework of the eye. It becomes the cornea at the front of the eye. The cornea is transparent, and with the lens makes up the focusing mechanism of the eye. In back of the cornea is a chamber containing a watery liquid called the aqueous humor. Covering the front of the sclera, except for the cornea, is the conjunctiva, which contains nerves that serve to alert one to foreign particles intruding under the eyelids.

The choroid comprises roughly 80% of the middle layer and is situated at the sides and rear of the eyeball. Most of the blood vessels

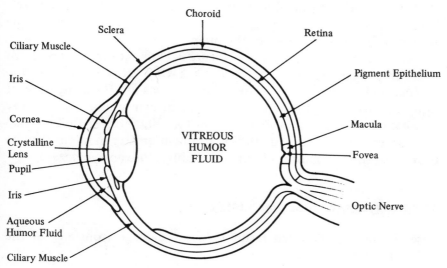

Figure 3-1. Schematic diagram of the human eye.

nourishing the eye are located in the choroid. This layer becomes the ciliary muscle, which through interconnecting fibers is attached to the crystalline lens. The ciliary muscles change the lens shape for focusing of the incoming light. Sandwiched between the lens and aqueous humor is the thin pigmented iris. The round black opening at the center of the iris is the pupil, an opening through which light is admitted. The muscles in the iris alter the quantity of light entering the eye by changing the size of the pupil.

The retina is the inner layer of the eyeball, being situated between the choroid and vitreous humor. A visual angle of approximately 240° in cross section is subtended by the retina. A retinal layer about 150 μm thick has nerve cells for light perception. Then comes the retinal pigment epithelium of 5 to 10 μm thickness, which serves to absorb scattered light and to stop light reflection. It also supports the photoreceptor cells. Two types of such light cells exist: the long, thin rods and the wider cones. Rods exhibit a strong sensitivity to low light levels but are incapable of discerning color. On the other hand, cones are capable of discerning color but are not as light-sensitive as rods. Cone density predominates near the center of the retina and rods near the periphery. Both have attaching nerve fibers that join together from all parts of the retina to form the optic nerve. This large nerve passes through the retina, choroid and sclera in the rear of the eyeball and goes to the brain.

The macula of the retina (see Figure 3-1) lies at the focal point of the cornea-lens system. The fovea is a tiny cross section of the macula and is the critical center for clear vision. Focusing of light onto the fovea by the cornea-lens system can amplify the energy density of the light by a factor of up to 100,000 over that striking the cornea. Since laser radiation is so much greater in intensity than normal light, the focusing effect can readily cause an eye hazard.

TISSUE DAMAGE MECHANISMS

Laser beam exposure can result in tissue damage from any of the following types of mechanisms:

- Thermal effects
- Photochemical effects

- Acoustic transients
- Chronic exposures
- Other phenomena

Thermal Effects

Thermally related effects represent the major type of damage from laser beam exposure. The extent of thermal damage is principally dependent upon the tissue that is the most efficient absorber of the particular laser radiation wavelength, the strength or irradiance of the incident laser beam, the size of the irradiated area and the exposure duration. Given the vast energy ranges of lasers, effects on tissue can vary from no effect, simple tissue reddening, steam generation, charring or even explosive tearing or ripping where very high radiant exposure levels occur.

Photochemical Effects

The photochemical effects on the eye and skin caused by ultraviolet radiation are not entirely comprehended. Nonetheless, action spectra graphs, which show the relative spectral effectiveness of actinic radiation (e.g., from the sun) in causing a skin reddening or erythema, have been experimentally defined.[3] Skin pigmentation plays a considerable role in establishing an erythema threshold.

ANSI[4] suggests that ultraviolet absorption by the corneal epithelium, and damage thereto, is probably caused by photochemical denaturization of molecules (e.g., DNA or RNA) or proteins in the cells.

Acoustic Transients

Laser pulses striking tissues have a part of their energy changed to a mechanical compression wave or acoustic transient. This acoustic energy is capable of ripping and tearing tissue when sufficiently intense. While the retinal injury mechanism is largely of thermal origin for Q-switched durations, the effect of acoustic transients caused by the rapid heating and thermal expansion in the proximity of the absorption site (e.g. each melanin granule) may play a part in the damage mechanism.[3]

Chronic Exposures

The damage mechanism and potential hazard to the eye or skin caused by chronic exposure to far-infrared ($\sim$3–1,000 μm) radiation is an area requiring further study. However, corneal exposures of 0.08 to 0.4 W · cm^{-2} in the near-infrared spectra emitted from foundry or glass furnaces are believed responsible for cataract formation.[5] Approximately 1 mW · cm^{-2} is the far-infrared corneal dosage received outdoors.

Chronic exposure to ultraviolet radiation is known to hasten aging of the skin, and it is believed that such exposure is responsible for causing certain skin cancers.[3]

Other Phenomena

Other possible damage mechanisms associated with still shorter pulse exposures than those of Q-switched lasers and mode-locked lasers include the following phenomena:[3]

- Direct electric-field effects
- Raman and Brillouin scattering
- Multi photon absorption

FACTORS CONTRIBUTING TO TISSUE INJURY

At least the following factors can play a role in causing tissue injury and influencing the degree of damage from laser beam exposures:

- Wavelength of laser radiation
- Tissue spectral absorption, reflection and transmission
- Strength or irradiance of incident laser beam
- Size of irradiated area
- Exposure duration
- Pupil size
- Location of retinal injury

Wavelength of Laser Radiation

Three major laser spectral bands are associated with biologic effects on tissues: visible, ultraviolet and infrared radiation wavelengths.

Frequent usage is made of spectral-band designations in discussing biologic effects. These follow:

$$\begin{aligned}
&\text{Near-ultraviolet:} && \text{UV-A (315–400 nm)} \\
& && \text{UV-B (280–315 nm)} \\
& && \text{UV-C (100–280 nm)} \\
&\text{Visible:} && \text{(400–700 nm)} \\
&\text{Near-infrared:} && \text{IR-A (700–1,400 nm)} \\
&\text{Far-infrared:} && \text{IR-B (1.4–3 } \mu\text{m)} \\
& && \text{IR-C (3–1,000 } \mu\text{m)}
\end{aligned}$$

Tissue Spectral Absorption, Reflection and Transmission

For the laser spectral bands of interest cited above, tissues will absorb, reflect and/or transmit the particular radiation. The various tissues behave differently at the different wavelengths. A cursory examination of behavior follows.

Visible and near-infrared radiation (400–1,400 nm) is transmitted through the cornea, aqueous humor, lens and vitreous humor. Significant quantities are absorbed at the retina and choroid. Figure

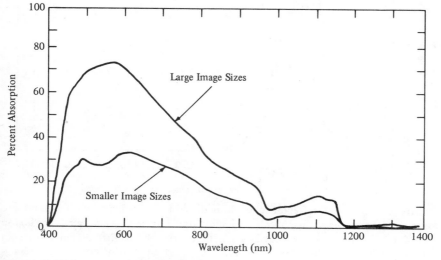

Figure 3-2. Spectral absorbed dose in retina and choroid relative to spectral corneal exposure as a function of wavelength. See David H. Sliney and Benjamin C. Freasier, *Applied Optics*, 12, 1 (1973).

3-2[3] shows the percent retinal and choroid absorption relative to corneal exposure as a function of wavelength.

Within the choroid and various retinal layers the behavior of the radiation is as follows:

- It is transmitted through the neural layers of the retina.
- The rods and cones absorb a small quantity of radiation.
- The rest of the energy gets absorbed in the choroid and pigment epithelium, the latter being optically the densest absorbent layer, which gives rise to the greatest temperature increase.

For incident infrared radiation of wavelength between 1,400 and 1,900 nm, the cornea and aqueous humor are almost total absorbers of the energy. At wavelengths greater than 1,900 nm the cornea is the sole absorber.

The crystalline lens of the eyeball is the principal absorber of the near-ultraviolet (315–400 nm) radiation. Insignificant radiation levels reach the retina.

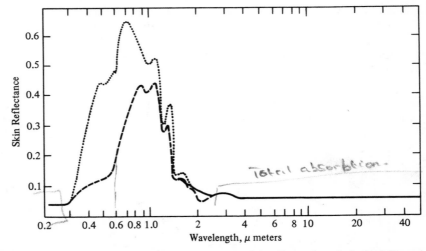

Figure 3-3. Spectral reflectance of human skin. Dotted and dashed lines are data of Jacquez and co-workers for individuals having a very fair complexion and for individuals having heavily pigmented skin. Spectral reflectance for skin in the ultraviolet and infrared regions reported by Hardy and co-workers is shown by a solid line. See David H. Sliney and Benjamin C. Freasier, *Applied Optics*, 12, 1 (1973).

As seen in Figure 3-3,[3] the spectral reflectance of human skin shows considerable variation in the UV-A, visible and IR-A bands. Also evident is the significant effect of skin pigmentation. Individuals with very fair complexions reflect more radiant energy and thus absorb less energy. On the other hand, persons with dark complexions, having heavily pigmented skin (more melanin granules), absorb more radiant energy and would be more prone to injury. As an example, at a wavelength of about 0.6 μm the skin reflectance is about 0.45 for individuals with very fair complexion, as against about 0.1 for individuals with heavily pigmented skin. Thus, irradiances would have to be increased for fair individuals to evoke the same response or injury.

It is also seen from Figure 3-3 that the spectral reflectance of the human skin is nearly constant at a value of about 0.05 for ultraviolet wavelengths under about 0.3 μm and for infrared wavelengths exceeding about 3 μm. Thus, almost total absorption occurs. For irradiated skin in the infrared region greater than about 3 μm, absorption takes place only in the most superficial layers.

Strength or Irradiance of Incident Laser Beam

It is intuitively clear that the greater the incident power density or irradiance impinging on tissue is, the greater the chance for damage.

It is of interest that high-powered infrared continuous wave (CW) output power lasers exceeding 1 KW are becoming quite common in certain applications. Yet, CW lasers with output exceeding 0.5 watt are classified as "high power" because a potential fire hazard and skin burn hazard exist.

Size of Irradiated Area

Tissue surrounding an absorption site is better able to conduct away the absorbed heat from a small rather than a large irradiated area. This strong dependence is illustrated in the following two examples:[3]

  1. Retinal injury thresholds for moderate exposure durations:
     - 1–10 W · cm^{-2} for a 1,000-μm image
     - Up to 1 KW · cm^{-2} for a 20-μm image

2. Sensation of warmth by human skin caused by incident beam of CO_2 laser emitting at 10.6 μm:
   - 0.1 W $\cdot$ cm^{-2} for a 1-cm-diameter image
   - 0.01 W $\cdot$ cm^{-2} for entire body exposure

Exposure Duration

The exposure level is simply the product of an irradiance (incident power density in W $\cdot$ cm^{-2}) and its duration exposure. Thus, it is clear that radiation exposure duration time is a fundamental parameter in accounting for tissue injury or in defining an injury threshold level.

Pupil Size

Pupil size can play a significant role in influencing retinal damage, since the energy admitted to the retina is proportional to the area of the pupil. An 8-mm pupil diameter permits 16 times the energy to fall upon the retina as does a 2-mm pupil.

Pupil size variations for various light environment conditions follow:

- 7–8 mm for the eye adapted to the dark
- 2–3 mm for outdoor daylight
- 1.6 mm for momentary view of the sun

Location of Retinal Injury

Inasmuch as various areas of the retina serve differently in the seeing process, the severity of injury is a function of the area effected. For example, a burn on the fovea, which is the area of sharpest vision, would cause a most drastic reduction of vision. Yet a similar-size burn on the peripheral retina could go totally undetected.

THE EYE HAZARD

Destruction of vision as a result of direct viewing of the sun has been known for a long time. While this handbook is principally concerned with the laser and associated hazards, and the laser is surely a serious hazard, it is worth noting that a number of other modern light sources exist that rival and exceed the sun's ability to cause eye damage. Compact arc lamps, electronic flash lamps, burning magnesium, as well as the nuclear fireball, are but a few such sources. The potential and extent of laser beam damage are generally dependent upon the particular structure of the eye that absorbs the most radiant energy per volume of tissue.

Table 3-1 summarizes many of the specific biologic effects on different structures of the eye, caused by laser radiation at the several spectral regions of interest.

THE SKIN HAZARD

Since laser radiation injuries to the eye can readily be of an irreversible nature, particularly where retinal injuries are involved, whereas skin damage is more readily reparable, the skin hazard is of distinctly secondary importance. However, with the increasingly widespread use of high energy lasers, the skin can sustain severe damage where energy densities approach several $J \cdot cm^{-2}$.

Like the eye, the skin is a body organ in that it handles important bodily tasks. Regulation of body heat is handled by the sweat glands and blood vessels, while oil gland secretion into hair follicles renders the hair smooth and glossy and prevents the skin from becoming too dry. The skin is one of the larger body organs, weighing about six pounds for an adult. Two major skin layers exist—the top layer called the epidermis and the underlayer termed the dermis. No blood vessels are present in the epidermis, but the lower layers contain some nerves. A pigment called melanin controls skin color, and is located in the lower epidermis layers. The greater the number of melanin granules is, the darker the skin color. The dermis varies from $\frac{1}{16}$ to $\frac{1}{8}$ inch in thickness, and contains sweat glands, blood vessels, vessels for moving the lymph, nerves, hair, hair oil glands and fat cells.

It is apparent that the skin is not a homogeneous organ. However, similar to most body tissue, the skin is comprised largely of water. Thus, it is relatively transparent to most laser radiation, at least up to the melanin pigment granules, which are the principal radiation absorbers of the skin. Visible, near-ultraviolet and infrared radiation

Table 3-1. Summary of Specific Biological Effects on the Eye.

Spectral Region	Absorption Area	Eye Schematic Representation	Specific Biological Effects	Comments
Far-Infrared (IR-B and IR-C) (1.4–1000 μm)	Cornea		• Minimal corneal lesion. • Loss of transparency of cornea or "glassworkers" or furnaceman's" cataract. • Increased irradiance can cause more serious damage.	A minimal corneal lesion is a small white area solely on the corneal epithelium, with the surface neither swollen nor elevated.[4] It appears within 10 minutes after exposure and heals within 48 hours without visible scars.
Short-Ultraviolet (UV-B and UV-C) (100–315 nm)	Cornea		• Excessive ultraviolet exposure can produce: – Redness – Tears – Secretion of mucus from the eyeball (conjunctival discharge) – Peeling or stripping off of the surface layer of cells from the cornea or connective tissue (stromal haze). • Damage to corneal epithelium by photochemical denaturization of proteins or other molecules (i.e., DNA, RNA, etc.).	Damage to the corneal epithelium is probably of photochemical rather than thermal origin.

Near-Ultraviolet (UV-A) (315–400 nm)	Primarily Lens		• Lens can flouresce. • Very high doses can cause corneal and lenticular opacities.
Light (400–700 nm) Near-Infrared (IR-A) (700–1,400 nm)	Retina and Choroid		• Mildest reaction may be simple reddening. • Minimal or threshold retinal lesion. • Increased retinal irradiance can cause large lesions, charring, hemorrhage, as well as gas formation, disruption of the retina and physical alteration of the eye structure.	A minimal retinal lesion is the smallest visible change in the retina viewed with an ophthalmoscope.[4] It occurs within a full day after exposure and is a small white patch (likely co-agulation).

Notes: (1) Compiled from References 4 and 6.

is selectively absorbed (see Figure 3-3) by the melanin granules. Sufficiently vigorous absorption under high energy density levels results in elevated temperatures and possible rupture of the granules. In general, skin damage can take the form of reddening, blistering, charring and actual burned-out cavities.

Photochemically active radiation or actinic radiation such as UV-B, C (100–315 nm) received from the sun accelerates skin aging with low-level chronic exposure. As noted earlier, it is also believed to cause several skin cancers. Some individuals have hypersensitivity reactions to radiation at specific spectral bands that result in skin reactions. This tendency or reaction is referred to as photosensitivity.

REFERENCES

1. "Eye," *The World Book Encyclopedia*, Vol. 6, Field Enterprises Educational Corporation, Chicago, pp. 357–360, 1968.
2. Van Pelt, W. F., Stewart, H. F., Peterson, R. W., Roberts, A. M., Worst, J. K., *Laser Fundamentals and Experiments*, U.S. Dept. of Health, Education and Welfare, Public Health Service, Bureau of Radiological Health, Southwestern Radiological Health Laboratory, Las Vegas, Nevada, pp. 28–35, May 1970.
3. Sliney, D. H., Freasier, B. C., "Evaluation of Optical Radiation Hazards," *Applied Optics*, Vol. 12, pp. 1–24, Jan. 1973.
4. "American National Standard for the Safe Use of Lasers, Z 136.1–1976," American National Standards Institute, New York, 1976.
5. Sliney, D. H., Vorpahl, K. W., Winburn, D. C., "Environmental Health Hazards From High-Powered, Infrared, Laser Devices," *Archives of Environmental Health*, Vol. 30, pp. 174–179, April 1975.
6. "Control of Hazards to Health from Laser Radiation," U.S. Dept. of the Army Technical Bulletin TB-MED-279, Washington, D.C., pp. 4, 5, May 1975.

4

Associated Laser Hazards

INTRODUCTION

The beam hazard is the single item of most concern to personnel working in the proximity of laser systems. However, the operation of laser systems like any other technological or industrial process involves various potential associated hazards. The following laser-associated hazards must be considered:

- Electrical
- Airborne contaminants
- Cryogenic liquids
- Noise
- Ionizing radiation
- Non–laser beam optical radiation
- Explosions
- Fire

Chapter 9 addresses the control measures to be used for each of the above items.

THE ELECTRICAL HAZARD

Franks and Sliney[1] have noted that laser-related accidents of an electrical origin are greater in number than eye and skin injuries from direct or reflected laser beams. Electrocution is a real possibility. Indeed, four documented electrocutions from laser-related activities have occurred in the United States.

Factors Influencing Severity of Electric Shock Injury

The degree of injury caused by electrical shock is influenced by at least the following factors:

- Current level
- Body resistance
- Current pathway through body
- Duration of shock
- Voltage
- Frequency of current
- Phase of heart cycle

The magnitude of the current flowing through the body is a key factor in determining electric shock injury. Various levels of electric current have been described and are defined below; some have been experimentally determined,[2,3] other values extrapolated from animal experiments,[3] with still others estimated. Table 4-1 provides some experimental and extrapolated levels.

- *Feeling threshold:* the current that will produce a slight sensation or tingling (see Table 4-1).
- *"Let-go" threshold:* that current level where volunteers tested are just able to let go of the conductor (see Table 4-1 for median "let-go" threshold values).
- *Freeze current:* that level which is slightly in excess of the "let-go" value, and which for AC voltage "freezes" the victim to the conductor.
- *Ventricular fibrillation currents:* moderate-size currents that can cause ventricular fibrillation of the heart (see Table 4-1).
- *Currents exceeding ventricular fibrillation levels:* moderate to very high current levels.

Table 4-1. Summary of Quantitative Effects of Electric Current on Humans.[1]

Effect	Direct Current (milliamperes)		Alternating Current (milliamperes)			
			60 Hertz[2]		10,000 Hertz	
	Men	Women	Men	Women	Men	Women
Slight sensation on hand	1	0.6	0.4	0.3	7	5
Median perception threshold	5.2	3.5	1.1	0.7	12	8
Not painful shock—muscular control not lost	9	6	1.8	1.2	17	11
Painful shock—muscular control lost by $\frac{1}{2}$% of volunteers	62	41	9	6	55	37
Painful shock—median "let-go" threshold	76	51	16	10.5	75	50
Painful and severe shock—breathing difficult, muscular control lost by $99\frac{1}{2}$% of volunteers	90	60	23	15	94	63
Possible ventricular fibrillation[3]						
• Three-second shocks	500	500	$\cong 100$	$\cong 100$		
• Short shocks (T in seconds)			$116/\sqrt{T}$	$116/\sqrt{T}$		
• Capacitor discharges	50*	50*				

*Energy in joules

[1] Adapted from Reference 3.
[2] Commercial voltage (e.g., 110 VAC).
[3] Estimates based on extrapolation from animal experiments.

The electrical resistance provided by the human body is made up of skin contact resistance and internal body resistance. Much variation exists for skin resistance values, with perspiration, skin puncture and conductor dampness all tending to lower the value. Dry uncut skin provides up to about 250,000 ohms of resistance. A 600-volt potential quickly breaks down the skin resistance, and essentially leaves a much lower internal body resistance of about 500 or so ohms for limiting the current through the body.

Where current passes through vital organs such as the heart, lungs or brain, the likelihood of serious injury increases greatly with increasing current. A current pathway from the hands to the feet, as is the case for many industrial accidents, involves the lungs and heart. This situation would be more serious than a foot-to-foot path.

The duration of the electrical shock can strongly effect the injury. For example, currents in excess of the "let-go" threshold for prolonged periods can render a victim insensible or unconscious or even cause death.

With the variability of body resistance, a definitive shock hazard voltage level is hard to define. High voltages are correctly recognized as hazardous. ANSI[4] considers levels over 42.5 volts as hazardous, if the current exceeds 0.5 milliamperes.

From Table 4-1 it is seen that 10 kHz frequency alternating current is a lesser hazard than 60 Hz current and direct current (DC). However, high voltage DC is more dangerous than high voltage alternating current.[5] Unlike alternating current, body contact with direct current does cause strong muscle contractions.

The susceptibility of the heart to ventricular fibrillation from electrical shocks during a portion of,[6] or the total,[1] partial refractory period of the heart cycle has been noted, and is thus a factor influencing electrical shock severity.

Physiological Effects of Electric Shock

The following physiological effects can occur when the body receives an electric shock:

- *Perception effects:* these occur at the body threshold level where a slight sensation or tingling is first experienced (see Table 4-1).

- *Some conscious phenomena:* as the current level is increased, pain and muscle soreness can occur, as well as temporary ringing in the ears and visual impairment such as flashes and spots before the eyes.
- *Muscular contractions:* alternating currents cause muscular contractions, with the severity dependent on the magnitude and frequency of the current. Levels at or below the "let-go" value (see Table 4-1), while painful, are of no lasting consequence; however, at the "freeze" value, where muscle contractions do not permit release of the conductor, the effects can be very serious if of a prolonged duration.
- *Blood pressure increase:* a rapid increase in blood pressure can occur as a result of strong muscle contractions.
- *Stupor, insensibility or collapse:* prolonged and/or strong shocks (e.g., at the "freeze" value) can render a person insensitive to the surroundings or cause collapse.
- *Unconsciousness:* prolonged and/or strong shocks at the "freeze" level can cause unconsciousness.
- *Paralysis of the breathing mechanism:* high voltages and/or currents passing through the chest or vital nerve centers can cause paralysis of the nerve centers, followed by asphyxiation and death.
- *Ventricular fibrillation:* with only moderate-size currents, ventricular fibrillation of the heart can occur, a condition where the rhythmic pumping action of the heart stops and blood circulation ceases. It is most hazardous, since heart fibrillation once started rarely reverts to normal rhythm naturally. A controlled countershock is required, and the needed equipment is not usually readily available. The shock levels shown in Table 4-1 are based on values extrapolated from animal experiments.
- *Burns:* two types of burns are of concern. First, are those caused by electric current flowing through the body tissues, which are slow to heal. Second are thermal burns caused by elevated temperatures near the body, such as from electric arcs, vaporized metal and other sources.
- *Hemorrhages:* depending on the magnitude of the shock, the effects can vary from simple blood shot eyes to severe hemorrhages of the brain, nervous system or other organs.

- *Heart standstill:* high currents can cause standstill of the heart, and death in a few minutes if artificial resuscitation is not applied promptly.
- *Nervous system damage:* relatively large currents can cause fatal damage to the central nervous system, with lesser currents providing nerve cell injury or temporary paralysis.
- *Increase in body temperature:* very high currents can raise the body temperature and produce almost instant death.

AIRBORNE CONTAMINANTS

A number of materials associated with laser technology and applications can give rise to potentially toxic atmospheric contaminants and concentrations so as to endanger the health of user personnel. Airborne contaminants include metallic fumes and dust, metallic oxide fumes, chemical and gaseous vapors and others. Biological contamination could be of concern where plume fragments from tissues spread living cancer tissue in the air after laser impact.[7]

Allowable Exposure Concentration

The Occupational Safety and Health Administration (OSHA)[8] cites two approaches to limiting contaminant concentrations so as to provide a safe environment for operating personnel. The first approach limits the employee's exposure to contaminant concentrations given by an "8-hour time weighted average" value, for any single 8-hour work day of a 40-hour work week. Second is an "acceptable ceiling concentration" value, which cannot be exceeded at any time in an 8-hour work day. For a small number of contaminants, the ceiling concentration level may be exceeded, up to a particular value, for periods varying from 5 to 30 minutes.

Table 4-2 shows the OSHA-prescribed limits for specific airborne contaminants arising from laser products and operations. OSHA does not provide allowable contaminant exposure levels for all potentially hazardous materials associated with lasers.

CRYOGENIC LIQUIDS

Some high-powered laser systems use cryogenic liquids to cool lasers and receiver sensor elements. Liquid nitrogen (LN 2) is a

Table 4-2. Allowable Airborne Contaminant Exposure Levels[1] for Some Laser Products and Operations.

Airborne Contaminant	Source of Contaminant	Allowable Contaminant Exposure Level[2,3]	
		8-Hour Time Weighted Average	Acceptable Ceiling Concentration
Asbestos	Target backstop	2 fibers (> 5 μm long)/cm^3	10 fibers (> 5 μm long)/cm^3
Bromine	Laser medium	0.1 ppm (0.7 mg/M^3)	—
Carbon dioxide	Laser medium	5,000 ppm (9,000 mg/M^3)	—
Carbon monoxide	Laser medium	50 ppm (55 mg/M^3)	—
Copper fumes	Metal target	0.1 mg/M^3	—
Hydrogen bromide	Laser discharge tube	3 ppm (10 mg/M^3)	—
Iodine	Ion laser medium	—	0.1 ppm (1 mg/M^3)
Iron oxide fume	Metal target	10 mg/M^3	—
Nickel	Metal target	1 mg/M^3	—
Nitrogen dioxide	Liquid nitrogen coolant	5 ppm (9 mg/M^3)	—
Ozone	Flash lamps	0.1 ppm	—
Selenium compounds	Liquid laser medium	0.2 mg/M^3	—
Tantalum	Wire target	5 mg/M^3	—
Zinc oxide fume	Target	5 mg/M^3	—

(1) Allowable exposure levels from OSHA standard of Reference 8.
(2) Parts of vapor or gas per million parts of contaminated air by volume at 25°C. and 760 mm Hg pressure.
(3) Approximate milligrams of particulate per cubic meter of air.

frequently used cryogenic, but liquid hydrogen and helium are also utilized.

The main hazards associated with the use of cryogenic fluids or cryogenic systems follow:

- *High pressure:* The large expansion ratios from cryogenic liquids to gas offer the potential for explosive build-up of pressures during the usage and storage of cryogenics.
- *Flammability:* Some gases, such as hydrogen, ozone and others, are inherently flammable. While oxygen is not flammable, its mixture with other materials and gases often increases flammability.
- *Bodily contact:* Short contact with cryogenic liquids can cause burns. Longer contacts may cause freezing and brittleness of the body part.
- *Asphyxiation:* Condensation (e.g., liquidification) of atmospheric oxygen can cause asphyxiation under confined and non-ventilated conditions. This is a possibility using liquid nitrogen, since oxygen condenses at about 13° Kelvin above liquid nitrogen.
- *Improper materials usage:* Properties of materials used for handling and storage of cryogenics must be carefully considered from a chemical reaction viewpoint and because of the low temperature effects (e.g., brittleness) on many materials. Thus, improper material selection causing decontainment of the cryogenic liquid is potentially hazardous.
- *Toxicity:* Some cryogenics such as carbon dioxide and fluoride, are inherently toxic and must be handled with great care.

The hazards are for the most part restricted to the research and development laboratories.

NOISE HAZARD

Discharging capacitors in the proximity of very high energy pulsed laser systems can sometimes produce hazardous noise levels. The Department of Labor[9] prohibits exposures to impact or impulse noise exceeding 140 dB peak sound pressure levels.

IONIZING RADIATION

Power supply tubes are capable of generating X-rays when potentials exceed 15 KV; however, most laser systems use voltages under 8,000 volts.[10] Input electrical voltages of 100 KV obtained using Marx generators are required for some high-powered gas lasers.[1]

NON-LASER BEAM OPTICAL RADIATION HAZARD

Apart from the laser beam itself, an ultraviolet radiation hazard is possible from optical pumping lamps and laser discharge tubes.

EXPLOSION HAZARD

The potential for explosions exists with faulty optical pumps and capacitor banks during the operation of high-powered laser systems. High pressure arc lamps and filament lamps used in laser equipment are likewise potential hazards. Care must be exercised in evaluating explosive or shatter effects of the laser target.

As previously noted, the storage and usage of cryogenics has a potential for explosive pressure build-up.

FIRE HAZARD

Laser-related fire hazards stem from either the direct laser beam if the energy is sufficiently high, or from electrical equipment. Components in electrical circuits must be evaluated with respect to fire hazards. Special consideration is required with power supply circuit wiring and transformers.

REFERENCES

1. Franks, J. F., Sliney, D. H., "Electrical Hazards of Lasers," *Electro-Optical Systems Design*, pp. 20–23, Dec. 1975.
2. Dalziel, C. F., "The Threshold of Perception Currents," *AIEE Annual Index to Electrical Engineering*, Vol. 73, pp. 625–630, New York, July 1954.
3. Dalziel, C. F., "Deleterious Effects of Electric Shock," *CRC Handbook of Laboratory Safety*, N. V. Steere, Editor, 2nd Edition, Chemical Rubber Co., Cleveland, Ohio, pp. 521–527, 1971.

4. "American National Standard for the Safe Use of Lasers, Z 136.1-1976," American National Standards Institute, New York, 1976.
5. Kouwenhoven, W. B., "Effects of Electricity on the Human Body," *AIEE Annual Index to Electrical Engineering*, Vol. 68, pp. 199–203, New York, March 1949.
6. Keesey, J. C., Letcher, F. S., "Human Thresholds of Electric Shock at Power Transmission Frequencies," *Archives of Environmental Health*, Vol. 21, pp. 547–552, Oct. 1970.
7. Goldman L., Rockwell, R. J., Jr., Hornby, P., "Laser Laboratory Design and Personnel Protection from High Energy Lasers," *CRC Handbook of Laboratory Safety*, N. V. Steere, Editor, 2nd Edition, Chemical Rubber Co., Cleveland, Ohio, pp. 381–389, 1971.
8. "Title 29, Code of Federal Regulations, Subpart G—Occupational Health and Environmental Control, Section 1910.93," Occupational Safety and Health Administration (OSHA) Standards, 29 Jan. 1974.
9. "Federal Register: Safety and Health Regulations for Construction," Vol. 36, No. 75, Part II, Section 1518.52, Bureau of Labor Standards, Dept. of Labor, 17 April 1971.
10. *Laser Safety Guide*, Third Edition, Laser Institute of America, Cincinnati, Ohio, Jan. 1976.

BIBLIOGRAPHY

1. Geiges, K. S., "Electric Shock Hazard Analysis," *AIEE Transactions Power Apparatus and Systems*, No. 23, pp. 1329–1331, Feb. 1957.
2. Price, L. D., "Codes and Standard Practices," *Standard Handbook for Electrical Engineers*, D. G. Fink, Editor, 10th Edition, Section 29, McGraw-Hill Book Co, New York, 1972.
3. Gagliano, F. P., "Lasers as a Thermal Energy source for Metal Working," *Engineering Publications*, University Park, Pa., pp. 46–63, Aug. 1970.
4. Ready, J. F., "Laser Applications in Metalworking," *Engineering Publications*, University Park, Pa., pp. 65–74, Aug. 1970.
5. *CRC Handbook of Laboratory Safety*, N. V. Steere, Editor, 2nd Edition, Chemical Rubber Co., Cleveland, Ohio, 1971.

5

Laser Measurements

INTRODUCTION

The measurements of principal interest in evaluating laser hazards are:

- Laser output energy and power
- Laser pulse width
- Irradiance and radiant exposure
- Beam distribution
- Beam divergence
- Pulse repetition rate

These parameters will be discussed, as will useful measurement devices and test techniques and precautions.

LASER OUTPUT ENERGY AND POWER

A number of techniques are used for measuring laser output. The use of calorimeters for this purpose has been highly successful, and frequently calorimeters are used as reference standards for other energy and power measurement devices such as quantum detectors.

Calorimetric Techniques

A fundamental method for measuring the output energy of a laser is to measure the temperature rise it causes in a material. If the material absorbs all the energy, there will be a temperature rise expressed by the equation

$$\Delta T = \frac{\Delta Q}{C(m)}$$

(Eq. 5-1)

where

ΔT = temperature rise in the material (°K)
m = mass of the material (gm)
C = specific heat of the material (in joules/gm-°K)
ΔQ = energy absorbed (joules)

One principal method for detecting the temperature rise in the material is to use a junction of dissimilar metals. Then a temperature rise in the junction will cause a voltage to develop across the junction. This is the typical action observed in thermocouples. Thermoelectric measurement of temperature can be very accurate. However, very small voltages must be measured. For example, the potential generated for a copper–constantan junction is 4 microvolts per degree Centigrade at 30°C.

Two other techniques of interest should be mentioned. First, if the material is a metal, the electrical resistance will increase as its temperature rises owing to exposure to the laser energy. The resistance change per °C is minute, and needs to be detected with a sensitive electrical bridge and galvanometer. A higher resistance change per °C can be obtained using a thermistor as the sensing device.

A second additional technique for detecting an increase in the temperature of a material is to measure expansion or change in the physical dimensions. Small amounts of radiant energy can be measured by using a fluid in a confined vessel and transferring the volumetric expansion into a linear change in a capillary tube.

Whatever technique is used, attention must be given to how much of the laser energy is reflected by the material, absorbed by the material, or transmitted through it. The percentage of energy absorbed must be accurately known. Even for the absorbed energy, heat losses can occur owing to radiation, convection, and conduction. Reemission of light by radiative transitions can also contribute to losses. Losses must be prevented or be controlled and of known quantity. Substitution heat calibration can be utilized to account for the above problems. An electrical resistance heater attached to the sensor material is used to introduce a known amount of heat. The output is correlated to the calibrated heat input.

Of the three approaches described above, the thermoelectric technique is probably the most widely used as an energy detector, despite its low microvolt output. Unlike bridge circuits, thermocouple techniques need not generate any heat at the measurement point.

Figure 5-1 illustrates the basic principle of a thermocouple laser energy meter. The input thermocouple junction receives the laser energy and undergoes a temperature change ΔT to T_1. The reference junction is at T_0. The meter gives an indication of the temperature difference $(T_1 - T_0)$. For zero temperature difference, the output is zero. This is another advantage of the thermocouple energy measurement technique.

Ordinarily, in a typical meter, an input aperture will pass ambient and laser energy to the sensing material, and an equivalent aperture will allow only ambient energy to fall on the referencing material. A number of thermocouples are positioned on the sensor material, with the same number in the reference portion of the meter. The two sets of thermocouples are connected so that their voltages have opposing

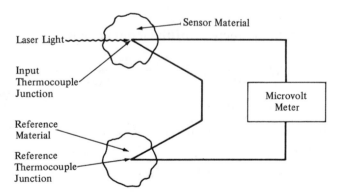

Figure 5-1. Basic principle of thermocouple laser energy meter.

polarities. If the two parts of the meter are at the same ambient temperature, the differential output potential will be zero. When a pulsed laser beam is supplied to the input, most of the energy will be absorbed, raising the temperature of the sensing material above that of the reference. The thermocouples on the input side generate a greater voltage, and a net potential will be measured at the output. The greater the number of thermocouples is, the greater the sensitivity. Cone-shaped collectors are frequently used to form the sensing and reference portions of the meter.

Example 5-1.   As an example, assume a need to calculate the sensitivity of a thermoelectric calorimeter. Assume that the cones are silver $(C = 0.056$ calorie/gm-$°$K), with a mass of 10 gm each. The thermocouples are made of Chromel-Alumel, and at a surrounding temperature of 30°C will generate a potential of 0.04 mV for each Kelvin degree temperature rise. Compute the output potential for an absorbed pulse of 1 joule. There are ten thermocouples on each cone.

1) Since there are about 4 joules/calorie, 1 joule of

$$\text{laser energy} = \Delta Q = \frac{1 \text{ joule}}{4 \text{ joules/calorie}} = 0.25 \text{ calorie}.$$

2) From Eq. (5-1):

$$\Delta T = \frac{\Delta Q}{C(m)}$$

The temperature increase of the input cone is:

$$\Delta T = \frac{0.25 \text{ calorie}}{(0.056 \text{ calorie/}°\text{K-gm}) (10 \text{ gm})}$$

$$T = 0.44°\text{K}$$

3) Each thermocouple on the reference cone will have an increased voltage output of:

$$(0.44°\text{K}) (0.04 \text{ mV/}°\text{K}) = 0.0176 \text{ mV}$$

4) The output will be the combined effect of all ten thermo-couples connected in series.

$$V_{\text{output}} = 10 \times (0.0176) = 0.176 \text{ mV}$$

Therefore, the sensitivity of the calorimeter will be about 176 microvolts per joule. Twenty thermocouples would give a sensitivity of 352 microvolts per joule.

To measure power with a calorimeter, the total heat is not usually measured; rather, the flow of heat from the sensor material is measured, with power indicated by a steady temperature drop along the heat exit path. A thermoelectric sensor that measures the flow of heat is called a heat flux sensor.[1] Simple and reliable laser calorimetric power and energy measuring devices have been developed by the National Bureau of Standards for both pulse and CW lasers. Numerous commercial models are also available.

Quantum Detectors

The basic fundamental calorimetric technique has been emphasized so far. However, photoemissive, photoconductive and photovoltaic detectors offer very good sensitivity in the 200–1,100 nm spectral region, and their use is widespread for laser energy and power meters.

The sensitivity of photoemissive types of detectors in a particular portion of the spectrum varies according to the photocathode material used in vacuum photodiodes or photomultiplier tubes. Silicon solid state photodiodes can function as either photovoltaic or photoconductive detectors.

Devices that have surfaces capable of ejecting an electron when a photon collides with the surface are called photoemissive. They can be used for measurement in the visible portion of the spectrum. The photomultiplier type of photoemissive device converts incident photons into a much greater, but proportional current. Radiant power can be calculated from the current, the photoemissive surface spectral sensitivity and the characteristics of any narrow band pass radiation filter used. Good sensitivity and fast response are advantages of photoemitters; however, their usefulness for power measurement is hampered by nonuniform spectral response. In addition, the photoemissive surfaces do not have uniform sensitivity.

Semiconductor P-N junction photodiodes of the photovoltaic type (solar cells) generate a voltage when photons are absorbed. The voltage can cause an external current to flow. Thus they are self-powered and can be used as simple, sensitive, linear power indicators.

Photoconductive semiconductors require a power supply for back-biasing. These semiconductors undergo changes in resistance when exposed to photons. They are capable of high-speed measurement of laser pulse shape.

Photographic Techniques

Photographic radiometry can be of value in laser power and energy measurements. Effective source size can be determined at different ranges. The beam distribution can also be evaluated. With sufficient care with respect to technique, radiant exposure (H) in joules per unit area can be estimated.

Pyroelectric Detectors

These detectors, which have been developed for commercial use in the last few years, are crystals that exhibit electrical effects when a temperature change is sensed. These effects can be used to detect and measure radiation.

The pyroelectric detector is a current source with an output that varies with the rate of change of the detector temperature. These detectors have extremely fast response and are insensitive to DC effects. They are now extensively utilized in radiometric systems, including widespread use in laser measurements.

LASER PULSE WIDTH

For a pulsed laser, once pulse energy is known, the power of a pulse can be obtained using the basic formula relating energy, power and time, with pulse width the principal factor involved.

$$\text{Energy} = \text{Power} \times \text{Time}$$

$$Q = \Phi \times t \qquad \text{(Eq. 5-2)}$$

$$\text{Joules} = \text{watts} \times \text{seconds}$$

Attention must be given to the shape of the pulse. Typically, an oscilloscope is used to monitor the pulse, receiving its input from a

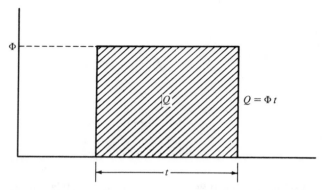

Figure 5-2. Square pulse energy-power-time relationship.

detector. For a simple, square or rectangular pulse, as in Figure 5-2, the relationship, $Q = \Phi t$, shows that energy is the area under the waveshape.

For the triangular (isosceles) pulse shown in Figure 5-3, the area under the curve is:

$$Q = \tfrac{1}{2} \Phi t$$

For the two examples shown Φ is the peak power. A general expression is shown in Figure 5-4:

$$Q = \int_{t_1}^{t_2} \Phi \, dt \qquad \text{(Eq. 5-3)}$$

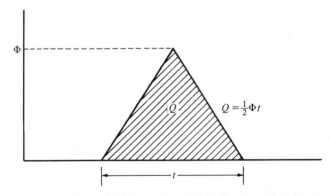

Figure 5-3. Triangular pulse energy-power-time relationship.

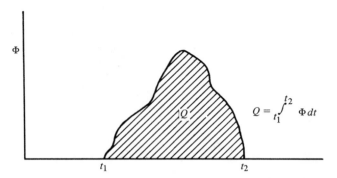

$$Q = \int_{t_1}^{t_2} \Phi \, dt$$

Figure 5-4. General pulse energy-power-time relationship.

In typical measurements, the pulses will bend at the bottom and top, but the triangular shape will often be a good approximation. See Figure 5-5, which shows the triangular shape approximating a Gaussian pulse, a pulse profile that many lasers approach. The curved pulse and the straight line intersect at about half the height of the waveforms. Pulse width is generally measured at these "half power" or half amplitude points.

Since the "half power" pulse width, $\tau_{1/2}$, is one-half the base of the pulse triangle, τ_B, we have:

$$Q = \tfrac{1}{2} \, \tau_B \Phi_{PK} = \tau_{1/2} \Phi_{PK}$$

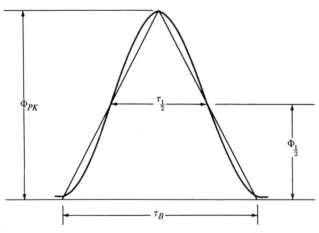

Figure 5-5. Gaussian pulse approximation with triangle.

To evaluate the output of a single pulse laser with a Gaussian waveform such as a Q-switched laser, proceed as follows:

- Measure the energy output, Q (with, for instance, a calorimeter).
- Record the pulse shape vs. time (with an oscilloscope and camera).
- Measure the pulsewidth, $\tau_{1/2}$.
- Compute the peak power, Φ_{PK}, using the formula $\Phi_{PK} = Q/\tau_{1/2}$.

IRRADIANCE AND RADIANT EXPOSURE

Two principal radiometric parameters required for hazard evaluation are irradiance, E, in $W \cdot cm^{-2}$ and radiant exposure, H, in $J \cdot cm^{-2}$. Another parameter used is radiance, L, in $W \cdot cm^{-2} \cdot sr^{-1}$. Additional information on these parameters and solved mathematical problems involving them can be found in Chapter 6. Table 5-1 summarizes the required radiometric parameters needed for determining eye and skin exposures from different laser sources for different wavelengths. Terms used in the table, such as extended sources, intrabeam source, and α_{min}, are discussed in Chapter 6.

Table 5-1. Required Radiometric Parameters.

	Eye			Skin
Laser Source	Visible (0.4–1.4 μm)	Ultraviolet (0.2–0.4 μm)	Infrared (1.4–10^3 μm)	All Wavelengths (0.2–10^3 μm)
Extended Sources (Angular subtense > α_{min})	Radiance	Irradiance	Irradiance	Irradiance
Intrabeam (Angular subtense < α_{min})	Irradiance	Irradiance	Irradiance	Irradiance

Source: American National Standards Institute, Reference 3.

BEAM DISTRIBUTION

Simplified approximation techniques have been given previously for evaluating Gaussian or normal pulse shapes, with the pulse varying in a normal manner vs. time.

The distribution of energy within a beam at any point in time is for many lasers also Gaussian. A Gaussian beam profile is characteris-

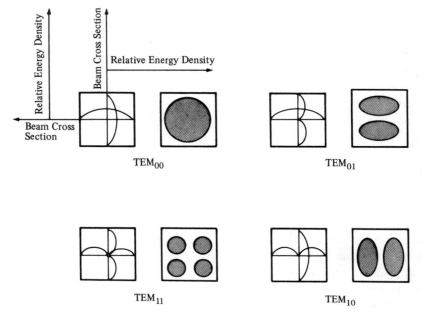

Figure 5-6. Beam cross sections for four different TEM modes.

tic of a single-mode laser. (See Figure 5-6 for some representations of different laser modes.) Figure 5-7 shows the irradiance cross section of a single-mode, Gaussian laser beam. If the beam diameter is defined as the $1/e$ points of the peak of the curve, the total beam power (or irradiance) divided by the area of a circular beam defined at these $1/e$ points gives a beam power equal to the peak shown in the curve. Figure 5-8 shows the relative power entering an aperture, relative to the beam diameter. The relative beam diameter is 1.0 at the $1/e$ points (marked a; the 87% or $1/e^2$ points are marked $a\sqrt{2}$).

Some laser manufacturers specify the beam diameter at $1/e^2$ of the peak power point. In this case, the total beam power (or irradiance) divided by the area of a circular beam defined at the $1/e^2$ points would yield only the average beam power; this could lead to misleading hazard evaluations.

Other manufacturers specify a diameter that contains 90% as compared to the 87% contained by the $1/e^2$ points. For safety purposes

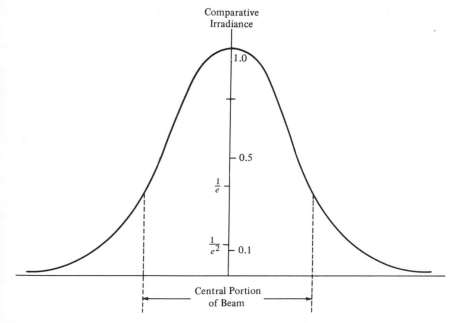

Figure 5-7. Irradiance cross section typical of a laser beam.

in determining Maximum Permissible Exposure (MPE), one must obtain the peak radiant exposure (H, in $J \cdot cm^{-2}$) or irradiance (E, in $W \cdot cm^{-2}$), for a wide beam diameter ($1/e$ diameter mentioned above), or the power or energy passing through a "limiting aperture." The limiting aperture is 7 mm, approximating a dilated pupil, for eye exposure limits in the visible and near-infrared (400–1,400 nm or 0.4–1.4 μm) wavelength region.[2] For ultraviolet wavelengths and much of the infrared wavelength region the limiting aperture is 1 mm, with the exception being a 1-cm aperture in the extreme infrared (0.1–1 mm) wavelength region.[1]

The beam profile is not always single-mode, as noted previously. The ruby laser is a typical example, the beam frequently breaking up into many nonsymmetrical parts. When a laser profile has no particular pattern, and perhaps also changes rapidly, the determination of a beam "diameter" is not readily made for safety hazard evaluation.

ANSI[3] indicates that the maximum radiance of an extended source

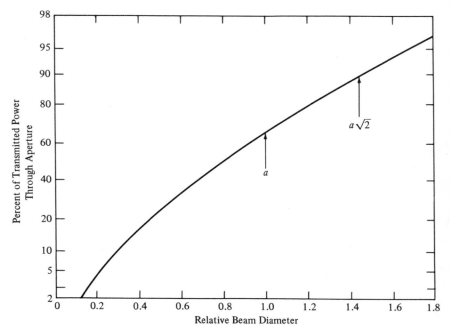

Figure 5-8. Percentage of the total laser beam power which passes through a circular aperture: Gaussian beam, single-mode laser.[2]

is to be determined, and that measurement may be averaged as follows:

- Over the conical field of view defined by the angular subtense, α_{min},
- Or else using a circular one-millimeter-diameter area, whichever yields the higher radiance value.

Where an extended source has a nonuniform profile (as could be caused by "hot spots"), the measurement is to be done for the area of highest radiance.

Table 5-2 is a rather detailed and recent table on apertures. The correct aperture is to be used in measuring and calculating values of irradiance and radiant exposure. The aperture is the largest circular area over which irradiance and radiant exposure are to be averaged.

Correction need not be made for beam size or nonuniformity when

Table 5-2. Maximum Aperture Diameters (Limiting Aperture) for Measurement Averaging.

Measurement	Exposure Duration, t (s)	Ultraviolet (0.2-0.4 μm)	Visible and Near-Infrared (0.4-1.4 μm)	Medium and Far-Infrared (1.4-10^2 μm)	Submillimeter (0.1-1 mm)
			Wavelength Range		
Eye MPE	10^{-9}-10^3	1 mm	7 mm	1 mm	11 mm
Skin MPE	10^{-9}-10^3	1 mm	1 mm	1 mm	11 mm
Laser Classification*	10^{-9}-10^3	80 mm	80 mm	80 mm	80 mm

*The apertures are used for the measurement of total output power or output energy for laser classification purposes, that is, to distinguish between all classes of cw lasers or between Class I and Class III pulsed lasers. The use of the 80-mm apertures as shown in the horizontal line labeled "Laser Classification" applies only to those cases where the laser output is intended to be viewed with optical instruments (excluding ordinary eyeglass lenses) or where the Laser Safety Officer determines that there is some probability that the output will be accidentally viewed with optical instruments and that such radiation will be viewed for a sufficient time duration so as to constitute a hazard. Otherwise the apertures listed for Eye MPE and Skin MPE are to be used.

For the specific case of optical viewing (beam collecting) instruments, the apertures listed for eye MPE and skin MPE apply to the exit beam of such devices.

Source: American National Standards Institute, Reference 3.

the whole beam passes through the limiting aperture. For wider beams, the hazard evaluation measurement is to be made in the portion of the beam that gives the highest reading.

BEAM DIVERGENCE

Basically, beam divergence can be measured by using a laser power meter at two distances separated by a distance r cm. Using a fixed aperture, the beam diameter in centimeters ($1/e$ point, for instance) is determined for each distance. After determining the change in beam diameter, Δd, in centimeters, beam divergence ϕ is calculated using

$$\phi = \frac{\Delta d}{r},$$
(Eq. 5-4)

where ϕ is in radians. Note that divergence is usually measured in milliradians.

PULSE REPETITION RATE

The best initial guide to pulse repetition rate is the information provided by the vendor or that indicated by switch settings where the rate is selectable. Oscilloscope measurements of the number of pulses for a given time base can be used when questions arise, backed up by photographs of the oscilloscope traces. Sensors feeding counter banks can also be used to implement an events-per-unit-time technique.

MEASUREMENT DEVICE EXAMPLES

A typical laser power measurement setup is schematically illustrated in Figure 5-9. A common type of laser, a 1.06 micron Neodymium-YAG variety, is the laser beam source. The particular instrument used is the EG&G 580/585. An energy collector, with a narrow-beam adapter, feeds the instrument. The output signal goes to a peak signal readout. Two calibration factors, k_1, for the 580/585, and k_2, for the energy collector, are required. The readout gives amperes per pulse, which when adjusted by $k_1 k_2$, gives a joules/pulse result. The instrument requires occasional periodic factory calibration. Spectral response is from 0.2 to 3.2 μm.

The basic radiometer is the Model 580 Radiometer. Pulse response is specified as under one nanosecond. Scope display provisions are included. EG&G lists the 580/585 as a spectroradiometer because of added optics and a grating monochromator that supplement the Model 580 Radiometer Detector Head.

Other typical devices include the Scientech Power Meter 364 and a 4″ diameter aperture absorbing calorimeter (also model 364), which gives a reading of joules per pulse. The Scientech calorimeter can be calibrated with an electrical power supply (substitution technique).

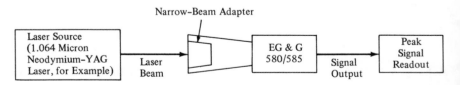

Narrow-Beam Adapter

Laser Source (1.064 Micron Neodymium-YAG Laser, for Example) — Laser Beam — EG & G 580/585 — Signal Output — Peak Signal Readout

Figure 5-9. Laser power measurement.

Molectron Corporation has a variety of instruments, including many using the pyroelectric effect for measuring laser power and energy.

REFERENCES

1. Zimmerer, R. W., "Theory and Practice of Thermoelectric Laser Power and Energy Measurement."
2. *Laser Safety Guide,* Third Edition, Laser Institute of America, Cincinnati, Ohio, Jan. 1976.
3. "American National Standards for the Safe Use of Lasers, Z 136.1-1976," American National Standards Institute, New York, 1976.

BIBLIOGRAPHY

1. Marshall, Samuel L., Editor, *Laser Technology and Applications*, McGraw-Hill Book Company, New York, 1968; particularly Chapter 8, "Laser Instrumentation," L. Waszak and T. Shultz.
2. Heard, H. G., *Laser Parameter Measurements Handbook*, John Wiley & Sons, Inc., New York, 1968.
3. Sliney, D. H., "Instrumentation and Measurement of Laser Radiation," AGARD Lecture Series No. 79, North Atlantic Treaty Organization, 1973.

6
Protective Standards

INTRODUCTION

Protective standards that provide criteria for exposure of the eye and skin to laser radiation contain values below known hazardous levels. These values are based on the best available information from experimental studies.

The most comprehensive protective standards presently available are those contained in the "American National Standard for the Safe Use of Lasers."[1] These standards, with only very minor modifications, are expected to provide the basis of user protective standards to be incorporated by the Occupational Safety and Health Administration (OSHA) of the HEW department. As such, the protective standards provided in this chapter are based on the ANSI (American National Standards Institute) information of Reference 1.

Maximum safe exposure levels are referred to by the ANSI, and by others, as Maximum Permissible Exposure or MPE levels, and the term shall be used in this handbook.

MPE determinations can become complex; for example, when several widely different laser wavelengths are emitted or when pulses are superimposed on a CW background. Numerous problems with their solutions (adapted from Reference 2) are included to assist the reader in acquiring skill in handling the MPE aspects of laser safety.

Major topics covered in this chapter are:

- Intrabeam and extended source exposure considerations
- Intrabeam viewing—Maximum Permissible Exposure (MPE)
- Extended source viewing—Maximum Permissible Exposure (MPE)
- Maximum Permissible Exposure (MPE) for skin exposure to a laser beam
- MPE correction factors and special handling techniques—visible and near infrared
- Determination of MPE for repetitively pulsed lasers
- MPE correction factors—infrared
- Formulas, considerations and examples useful in evaluation of various laser applications

INTRABEAM AND EXTENDED SOURCE EXPOSURE CONSIDERATIONS

Exposure of the eye to a laser beam can occur in one of two manners, namely, as a result of intrabeam viewing or of extended source viewing. Appropriate protective standards or MPEs are associated with each viewing case.

Simple pictorial illustrations of intrabeam and extended source viewing follow, along with mathematical symbol definitions used throughout this chapter for formula and problem development.

Diagrams and Symbols

Figures 6-1, 6-2 and 6-3, as well as the following symbol definitions, were adopted from Reference 2.

a = Diameter of emergent laser beam (cm).
d_e = Diameter of the pupil of the eye (varies from approximately 0.2 to 0.7 cm).
d_{min} = Limiting object size of extended object (cm).
D_e = Diameter of the exit pupil of an optical system (cm).
D_L = Diameter of laser beam at range r (cm).

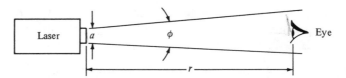

Figure 6-1. Intrabeam viewing—direct primary beam.

D_o = Diameter of objective of an optical system (cm).
e = Base of natural logarithms.
E, H = Radiant exposure (H) or irradiance (E) at range r, measured in $J \cdot cm^{-2}$ for pulsed lasers and $W \cdot cm^{-2}$ for CW lasers.
E_0, H_0 = Emergent beam radiant exposure (H_0) or irradiance (E_0) at zero range (units as for E, H).
f = Effective focal length of eye (1.7 cm).
F = Pulse repetition frequency (PRF), s^{-1} or Hz.
G = Ratio of retinal irradiance or radiant exposure received by optically aided eye to that received by unaided eyes.
L = Radiance of an extended source ($W \cdot cm^{-2} \cdot sr^{-1}$).
L_p = Integrated radiance of an extended source ($J \cdot cm^{-2} \cdot sr^{-1}$).
P = Magnifying power of an optical system.
Q = Total radiant energy output of a pulsed laser, measured in joules.
r = Range from the laser to the viewer or to a diffuse target (cm).
r_1 = Range from the laser target to the viewer (cm).
r_{1max} = Maximum range from the laser target to the viewer where extended source protection standard (MPE) applies (cm).
r_a = Effective range from the curved laser target to the viewer (cm).
R = Radius of curvature of a specular surface (cm).
S = Scan rate of a scanning laser (number of scans across eye per second).
T = Total exposure duration (in seconds) of a train of pulses.
T_i = Integrated "on-time" of a train of pulses (TOTP).
t = Duration of single pulse(s).
T_{max} = Classification duration, i.e., a maximum duration of daily exposure inherent in the design of the laser device.

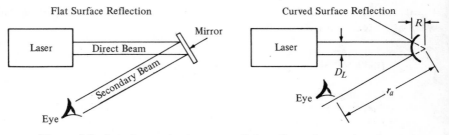

Figure 6-2. Intrabeam viewing—specularly reflected secondary beam.

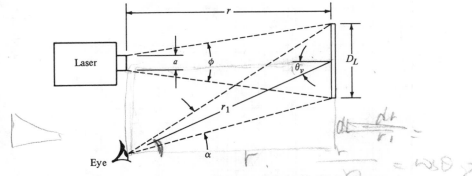

Figure 6-3. Extended source viewing—normally diffuse reflection.

α = Viewing angle subtended by an extended source (radians).
α_{min} = Minimum angle subtended by a source for which extended source MPE applies (radians).
μ = Atmospheric attenuation coefficient (cm^{-1}) at a particular wavelength.
ϕ = Emergent beam divergence (radians).
Φ = Total radiant power (or radiant flux) output of a CW laser, or average radiant power of a repetitively pulsed laser, measured in watts.
ρ_λ = spectral reflectance of a diffuse object at wavelength λ.
τ = Transmittance of a filter.
θ_s = Maximum angular sweep of a scanning beam (radians).
θ_v = Viewing angle (see Figure 6-3).

The Limiting Angle (α_{min})

The angle α_{min} is used to check whether the viewing range (r_1) for a given circumstance is close enough to apply extended source protection standards. Figure 6-3 shows the geometric relationship of r_1, D_L, θ_v and α_{min}. From the general case,

$$\alpha = \frac{D_L \cdot \cos \theta_v}{r_1} \qquad \text{(Eq. 6-1)[1,2]}$$

it follows that

$$\alpha_{min} = \frac{D_L \cdot \cos \theta_v}{r_{1max}} \qquad \text{(Eq. 6-2)[1,2]}$$

for $\theta_v \leq 0.37$ radian ($21°$).

The protection standards for extended sources apply to sources subtending an angle equal to or greater than α_{min} of Figure 6-4, and

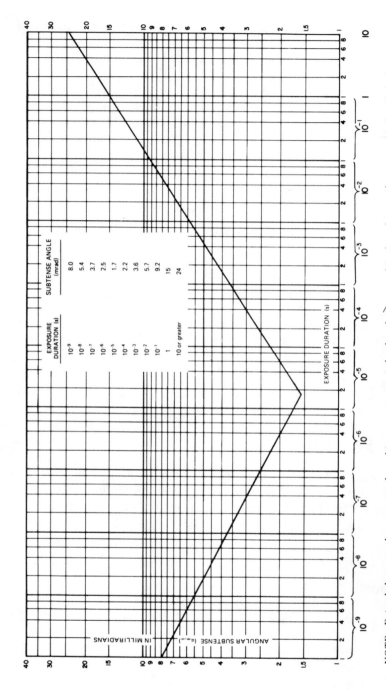

NOTE: Extended sources have an angular subtense or apparent visual angle $\geqq \alpha_{min}$. Angular subtenses (apparent visual angles) $< \alpha_{min}$ are considered intrabeam viewing.

Figure 6-4. Limiting angular subtense (apparent visual angle), α_{min}. Source: American National Standards Institute, Reference 1.

varies with exposure time. For angles less than α_{min}, intrabeam viewing MPEs apply. Thus, the limiting angular subtense is that apparent visual angle which separates intrabeam viewing from extended source viewing.

Intrabeam Viewing Sources

Some intrabeam viewing sources follow:

- Direct collimated laser beam capable of generating small retinal images.
- Flat or curved specular surfaces that reflect collimated laser beam, which is capable of generating small retinal images.
- Diffuse reflecting surface struck by a collimated laser beam, or energy from laser arrays or diodes, for any of the images where the angular subtense is less than α_{min} of Figure 6-4.

Extended Source Viewing

Diffuse reflecting surfaces struck by a laser beam, or energy from laser arrays or diodes, may act as extended sources for any of the images whose angular subtense is equal to or greater than α_{min} of Figure 6-4.

INTRABEAM VIEWING—MAXIMUM PERMISSIBLE EXPOSURE (MPE)

Protective standards or MPE values for direct ocular exposure to single pulses or exposures for intrabeam viewing are provided in Table 6-1. Figures 6-5, 6-6 and 6-7 supplement the Table 6-1 information by providing graphical presentations of the protective standards.

Various correction factors to the MPE values are covered in a later section of this chapter.

Example 6-1 is included for determination of the MPE for a single laser pulse intrabeam viewing case.

Table 6-1. Maximum Permissible Exposure (MPE) for Direct Ocular Exposures. Intrabeam Viewing from a Laser Beam.

Wavelength, λ (μm)	Exposure Duration, t (s)	Maximum Permissible Exposure (MPE)	Notes for Calculation and Measurement
Ultraviolet			
0.200–0.302	10^{-2}–3×10^4	3×10^{-3} J·cm^{-2}	
0.303	10^{-2}–3×10^4	4×10^{-3} J·cm^{-2}	
0.304	10^{-2}–3×10^4	6×10^{-3} J·cm^{-2}	
0.305	10^{-2}–3×10^4	1.0×10^{-2} J·cm^{-2}	
0.306	10^{-2}–3×10^4	1.6×10^{-2} J·cm^{-2}	
0.307	10^{-2}–3×10^4	2.5×10^{-2} J·cm^{-2}	1-mm limiting aperture
0.308	10^{-2}–3×10^4	4.0×10^{-2} J·cm^{-2}	In no case shall the total irradiance, over all
0.309	10^{-2}–3×10^4	6.3×10^{-2} J·cm^{-2}	the wavelengths within the UV spectral
0.310	10^{-2}–3×10^4	1.0×10^{-1} J·cm^{-2}	region, be greater than 1 watt per square
0.311	10^{-2}–3×10^4	1.6×10^{-1} J·cm^{-2}	centimeter upon the cornea
0.312	10^{-2}–3×10^4	2.5×10^{-1} J·cm^{-2}	
0.313	10^{-2}–3×10^4	4.0×10^{-1} J·cm^{-2}	
0.314	10^{-2}–3×10^4	6.3×10^{-1} J·cm^{-2}	
0.315–0.400	10^{-9}–10	$0.56\, t^{1/4}$ J·cm^{-2}	
0.315–0.400	10–10^3	1 J·cm^{-2}	
0.315–0.400	10^3–3×10^4	1×10^{-3} W·cm^{-2}	

Visible and Near-Infrared*

0.400–0.700	10^{-9}–1.8×10^{-5}	5×10^{-7} J $\cdot$ cm^{-2}
0.400–0.700	1.8×10^{-5}–10	$1.8\, t^{3/4} \times 10^{-3}$ J $\cdot$ cm^{-2}
0.400–0.550	10–10^{4}	10×10^{-3} J $\cdot$ cm^{-2}
0.550–0.700	10–T_1	$1.8\, t^{3/4} \times 10^{-3}$ J $\cdot$ cm^{-2}
0.550–0.700	T_1–10^{4}	$10\, C_B \times 10^{-3}$ J $\cdot$ cm^{-2}
0.400–0.700	10^{4}–3×10^{4}	$C_B \times 10^{-6}$ W $\cdot$ cm^{-2}
0.700–1.059	10^{-9}–1.8×10^{-5}	$5\, C_A \times 10^{-7}$ J $\cdot$ cm^{-2}
0.700–1.059	1.8×10^{-5}–10^{3}	$1.8\, C_A\, t^{3/4} \times 10^{-3}$ J $\cdot$ cm^{-2}
1.060–1.400	10^{-9}–5×10^{-5}	5×10^{-6} J $\cdot$ cm^{-2}
1.060–1.400	5×10^{-5}–10^{3}	$9\, t^{3/4} \times 10^{-3}$ J $\cdot$ cm^{-2}
0.700–1.400	10^{3}–3×10^{4}	$320\, C_A \times 10^{-6}$ W $\cdot$ cm^{-2}

} 7-mm limiting aperture
See Table 6-5 for correction factors

Far-Infrared

1.4–10^{3}	10^{-9}–10^{-7}	10^{-2} J $\cdot$ cm^{-2}
	10^{-7}–10	$0.56\, t^{1/4}$ J $\cdot$ cm^{-2}
	>10	0.1 W $\cdot$ cm^{-2}

} See Table 5-2 for apertures
See page 73 for correction factors and Fig. 6-7.

*See Fig. 6-5 for graphic presentation.

NOTES

$C_A = \text{CF}_{8.5.2}$ of Figure 6-9
$C_B = 1$ for $\lambda = 0.400 - 0.550$ μm (see Fig. 6-10)
$C_B = 10^{[15(\lambda - 0.550)]}$ for $\lambda = 0.550 - 0.700$ μm
$T_1 = 10$ s for $\lambda = 0.400 - 0.550$ μm (see Fig. 6-10)
$T_1 = 10 \times 10^{[20(\lambda - 0.550)]}$ for $\lambda = 0.550 - 0.700$ μm
Source: American National Standards Institute, Reference 1.

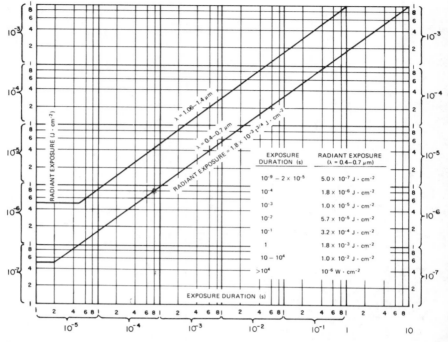

NOTE: For correction factors at wavelengths between 0.7 and 1.4 μm, see Table 6-5.

Figure 6-5. MPE for direct ocular exposure to visible radiation (λ = 0.4–1.4 μm), intrabeam viewing,* for single pulses or exposures. Source: American National Standards Institute, Reference 1.

*Angular subtense less than α_{min} in Fig. 6-4.

Example 6-1. Single Pulse Visible Laser. Find the protective standard (MPE) for a direct intrabeam exposure to a 694.3-nm ruby laser pulse that has a duration of 0.8 millisecond.

1)   λ = 694.3 nm

t = 0.8 millisecond = 8 X 10⁻⁴ second

2) Method #1
   Read value directly from Figure 6-5. For a single pulse exposure duration of 8 X 10⁻⁴ second, the radiant exposure (H) is approximately 8 X 10⁻⁶ J · cm⁻².

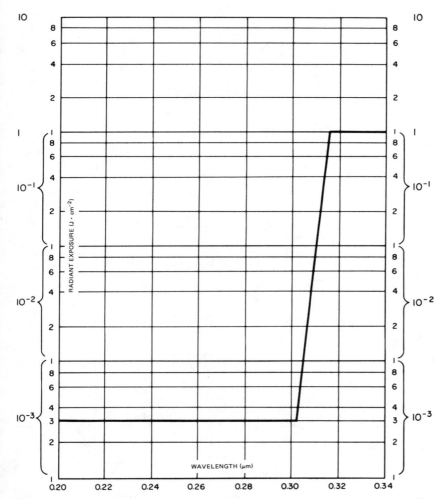

Figure 6-6. MPE for direct ocular exposure to ultraviolet radiation for exposure durations from 10^{-2} to 3×10^{4} second. Source: American National Standards Institute, Reference 1.

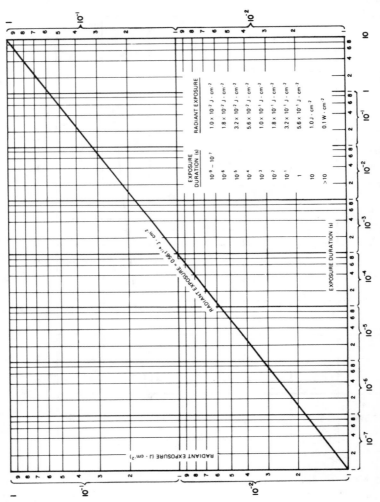

NOTE: For correction factors at 1.54 μm, see page 73.

Figure 6-7. MPE for ocular exposure to infrared radiation (λ = 1.4 μm–1 mm) for single pulses or exposures. Source: American National Standards Institute, Reference 1.

3) Method #2

- From Table 6-1, the MPE for a laser exposure of 8×10^{-4} second (i.e., between 1.8×10^{-5} and 10 seconds) is

$$\text{MPE: } H = 1.8 \, t^{3/4} \times 10^{-3} \text{ J} \cdot \text{cm}^{-2}$$
$$= 1.8 \, (8 \times 10^{-4})^{3/4} \times 10^{-3} \text{ J} \cdot \text{cm}^{-2}$$
$$= 8.56 \times 10^{-6} \text{ J} \cdot \text{cm}^{-2}$$

- Since $E \cdot t = H$, the MPE can also be in terms of irradiance (E). Thus,

$$\text{MPE: } E = \frac{H}{t} = \frac{8.56 \times 10^{-6} \text{ J} \cdot \text{cm}^{-2}}{8 \times 10^{-4} \text{ second}}$$
$$= 1.07 \times 10^{-2} \text{ W} \cdot \text{cm}^{-2}$$

EXTENDED SOURCE VIEWING—MAXIMUM PERMISSIBLE EXPOSURE (MPE)

Protective standards or MPE values for ocular exposure to extended sources for single pulses or exposures are provided in Table 6-2. Note that all values are specified at the cornea. Figures 6-7 and 6-8 supplement the Table 6-2 information by providing graphical presentations of the protective standards.

Various correction factors to the MPE values are covered in a later section of this chapter.

Examples follow for determination of the MPE for single laser pulse extended source viewing cases.

Example 6-2. Finding the Maximum Distance Where The Extended Source MPE Applies. Determine the maximum distance r_1 for a visible laser that has an emergent beam diameter $a = 1$ cm, a beam divergence $\phi = 10^{-4}$ radian and a pulse duration of 20 μsec. A diffuse target is placed one meter from the beam exit of the laser (e.g., target distance $r = 1$ meter).

1)
$$a = 1 \text{ cm}$$
$$\phi = 10^{-4} \text{ radian}$$
$$t = 20 \, \mu\text{sec} = 2 \times 10^{-5} \text{ second}$$
$$r = 1 \text{ meter} = 100 \text{ cm}$$

Table 6-2. Maximum Permissible Exposure (MPE) for Viewing a Diffuse Reflection of a Laser Beam or an Extended Source Laser.

Wavelength, λ (μm)	Exposure Duration, t (s)	Maximum Permissible Exposure (MPE)	Notes for Calculation and Measurement
Ultraviolet			
0.200–0.302	10^{-2}–3×10^4	3×10^{-3} J·cm^{-2}	
0.303	10^{-2}–3×10^4	4×10^{-3} J·cm^{-2}	
0.304	10^{-2}–3×10^4	6×10^{-3} J·cm^{-2}	
0.305	10^{-2}–3×10^4	1.0×10^{-2} J·cm^{-2}	
0.306	10^{-2}–3×10^4	1.6×10^{-2} J·cm^{-2}	
0.307	10^{-2}–3×10^4	2.5×10^{-2} J·cm^{-2}	1-mm limiting aperture
0.308	10^{-2}–3×10^4	4.0×10^{-2} J·cm^{-2}	
0.309	10^{-2}–3×10^4	6.3×10^{-2} J·cm^{-2}	
0.310	10^{-2}–3×10^4	1.0×10^{-1} J·cm^{-2}	
0.311	10^{-2}–3×10^4	1.6×10^{-1} J·cm^{-2}	
0.312	10^{-2}–3×10^4	2.5×10^{-1} J·cm^{-2}	
0.313	10^{-2}–3×10^4	4.0×10^{-1} J·cm^{-2}	
0.314	10^{-2}–3×10^4	6.3×10^{-1} J·cm^{-2}	
0.315–0.400	10^{-9}–10	$0.56 \, t^{1/4}$ J·cm^{-2}	
0.315–0.400	10–10^3	1 J·cm^{-2}	
0.315–0.400	10^3–3×10^4	1×10^{-3} W·cm^{-2}	

Visible*

0.400–0.700	10^{-9}–10	$10\,t^{1/3}$ J · cm^{-2} · sr^{-1}	
0.400–0.550	10–10^4	21 J · cm^{-2} · sr^{-1}	1-mm limiting aperture or α_{min}, whichever is greater
0.550–0.700	10–T_1	$3.83\,t^{3/4}$ J · cm^{-2} · sr^{-1}	
0.550–0.700	T_1–10^4	$21\,C_B$ J · cm^{-2} · sr^{-1}	See Table 6-5 for correction factors and Fig. 6-8
0.400–0.700	10^4–3×10^4	$2.1\,C_B\,t\,10^{-3}$ W · cm^{-2} · sr^{-1}	

Near Infrared*

0.700–1.400	10^{-9}–10	$10\,C_A\,t^{1/3}$ J · cm^{-2} · sr^{-1}	1-mm limiting aperture or α_{min}, whichever is greater
0.700–1.400	10–10^3	$3.83\,C_A\,t^{3/4}$ J · cm^{-2} · sr^{-1}	
0.700–1.400	10^3–3×10^4	$0.64\,C_A$ W · cm^{-2} · sr^{-1}	See Table 6-5 for correction factors and Fig. 6-8

Far-Infrared

1.4–10^3	10^{-9}–10^{-7}	10^{-2} J · cm^{-2}	See Table 5-2 for apertures
	10^{-7}–10	$0.56\,t^{1/4}$ J · cm^{-2}	See page 73 for correction factors and Fig. 6-7.
	>10	0.1 W · cm^{-2}	

*See Fig. 6-8 for graphic presentation and Fig. 6-3.

NOTES

$C_A = \mathrm{CF}_{8.5.2}$ of Fig 6-9

$C_B = 1$ for $\lambda = 0.400 - 0.550$ μm (see Fig. 6-10)

$C_B = 10^{[15(\lambda - 0.550)]}$ for $\lambda = 0.550 - 0.700$ μm

$T_1 = 10$ s for $\lambda = 0.400 - 0.550$ μm (see Fig. 6-10)

$T_1 = 10 \times 10^{[20(\lambda - 0.550)]}$ for $\lambda = 0.550 - 0.700$ μm

Source: American National Standards Institute, Reference 1.

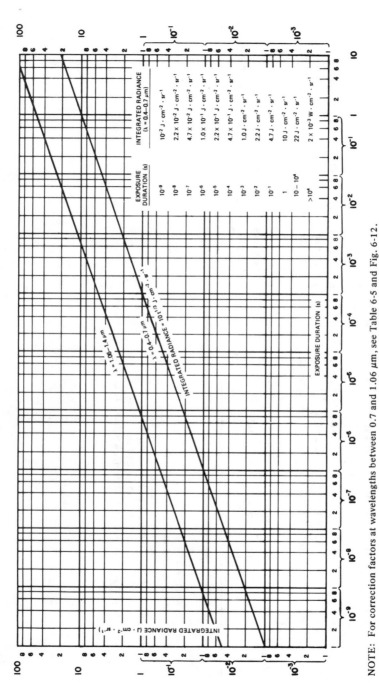

NOTE: For correction factors at wavelengths between 0.7 and 1.06 μm, see Table 6-5 and Fig. 6-12.

Figure 6-8. MPE for direct ocular exposure to laser radiation (λ = 0.4–0.7 μm and 1.06–1.4 μm) from extended sources* for single pulses or exposures. Source: American National Standards Institute, Reference 1.

*Angular subtense $\geq \alpha_{min}$ in Fig. 6-4.

2) From Figure 6-3, it can be shown that the relation of D_L to the emergent beam divergence and diameter is given by

$$D_L = a + r \cdot \phi \qquad \text{(Eq. 6-3)}$$

3) At short target distances, D_L is very nearly equal to a. For example, with $r = 100$ cm and $\phi = 10^{-4}$ radian, then from Equation 6-3,

$$D_L = 1 + (100)(10^{-4}) = 1.01 \text{ cm}$$

$$\approx 1 \text{ cm}$$

4) From Figure 6-4, with exposure duration $t = 2 \times 10^{-5}$ second,

$$\alpha_{min} = 1.6 \text{ millradians}$$

$$= 1.6 \times 10^{-3} \text{ radian}$$

5) From Eq. 6-2, and assuming worst-case viewing condition $(\theta_v = 0°)$,

$$r_{1max} = \frac{D_L \cdot \cos \theta_v}{\alpha_{min}} = \frac{(1 \text{ cm})(1)}{1.6 \times 10^{-3} \text{ radian}}$$

$$= 625 \text{ cm}$$

6) For distances less than 625 cm, extended source viewing applies. From Figure 6-8, with an exposure duration of $t = 2 \times 10^{-5}$ second, for visible laser λ falls between 400 and 700 nm, the applicable

$$\text{MPE} \cong 2.8 \times 10^{-1} \text{ J} \cdot \text{cm}^{-2} \cdot \text{sr}^{-1}$$

7) At distances greater than 625 cm, intrabeam viewing applies. Again, λ falls between 400 and 700 nm, and with an exposure duration of $t = 2 \times 10^{-5}$ second, from Figure 6-5, the applicable

$$\text{MPE} = 5 \times 10^{-7} \text{ J} \cdot \text{cm}^{-2}$$

Example 6-3. Extended Source MPEs for Diffuse Reflections Expressed as Incident Beam Irradiance (H) or Radiant Exposure (E). Consider the laser of Example 6-2 being fired at a 100% reflectance white diffuse target. Find the MPE expressed in terms of incident beam irradiance (H).

Table 6-3. Typical Diffuse Reflection Cases for Visible Radiation (λ = 0.4-0.7 μm).*

| Exposure Duration, t (s) | MPE at Cornea† ($J \cdot sr^{-1} \cdot cm^{-2}$) | Limiting Angle‡, α_{min} | Permissible Laser Beam Radiant Exposure Incident on the Diffusely Reflecting Surfaces, MPE $\times \pi/\rho$ ($J \cdot cm^{-2}$) | | |
| | | | Reflectance (ρ) = 100% | Reflectance (ρ) = 50% | Reflectance (ρ) = 10% |
(1)	(2)	(3)	(4)	(5)	(6)
10^{-9}	1.0×10^{-2}	8.0	3.1×10^{-2}	6.3×10^{-2}	3.1×10^{-1}
10^{-8}	2.2×10^{-2}	5.4	6.8×10^{-2}	1.4×10^{-1}	6.8×10^{-1}
10^{-7}	4.6×10^{-2}	3.7	1.5×10^{-1}	2.9×10^{-1}	1.5
10^{-6}	1.0×10^{-1}	2.5	3.1×10^{-1}	6.3×10^{-1}	3.1
10^{-5}	2.2×10^{-1}	1.7	6.8×10^{-1}	1.4	6.8
10^{-4}	4.6×10^{-1}	2.2	1.5	2.9	15
10^{-3}	1.0	3.6	3.1	6.3	31
10^{-2}	2.2	5.7	6.8	14	68
10^{-1}	4.6	9.2	15	29	150
1	10	15	31	63	310
$10-10^4$	22	24	68	140	680

*For wavelengths between 0.7 and 1.4 μm, use the appropriate correction factors given in Table 6-5.

†See Fig. 6-8 for a graphic presentation of these values.

‡For extended sources, angular subtense $\geq \alpha_{min}$.

Source: American National Standards Institute, Reference 1.

1) Spectral reflectance $\rho_\lambda = 100\% = 1.0$.
   Integrated radiance $L_p = 2.8 \times 10^{-1}$ J $\cdot$ cm^{-2} $\cdot$ sr^{-1} (from step 6 of Example 6-2).

2)
$$L_p = \frac{H \cdot \rho_\lambda}{\pi}$$
(Eq. 6-4)[1,2]

   or

$$H = \frac{\pi L_p}{\rho_\lambda}$$
(Eq. 6-5)[1,2]

$$= \frac{(3.14)\,(2.8 \times 10^{-1} \text{ J} \cdot \text{cm}^{-2} \cdot \text{sr}^{-1})}{1.0}$$

$$= 0.88 \text{ J} \cdot \text{cm}^{-2}$$

3) Alternate Approach

   • From Example 6-2, exposure duration $t = 2 \times 10^{-5}$ second.
   • Using Table 6-3 and interpolating the values in column 4 between 10^{-5} and 10^{-4} second, we get:

$$\frac{6.8 \times 10^{-1} \text{ J} \cdot \text{cm}^{-2}}{H} = \frac{2.2 \times 10^{-1} \text{ (J} \cdot \text{cm}^{-2} \cdot \text{sr}^{-1})}{2.8 \times 10^{-1} \text{ (J} \cdot \text{cm}^{-2} \cdot \text{sr}^{-1})}$$

$$H = 0.87 \text{ J} \cdot \text{cm}^{-2}$$

MAXIMUM PERMISSIBLE EXPOSURE (MPE) FOR SKIN EXPOSURE TO A LASER BEAM

Protective standards or MPE values for skin exposure to a laser beam are given in Table 6-4.

Various correction factors to the MPE values are covered in the following section of this chapter.

MPE CORRECTION FACTORS AND SPECIAL HANDLING TECHNIQUES—VISIBLE AND NEAR INFRARED

A number of wavelength and exposure duration–dependent corrections, as well as special handling techniques for multiple laser pulses, are needed for a proper determination of the MPE, and are summarized in Table 6-5.

Table 6-4. Maximum Permissible Exposure (MPE) for Skin Exposure to a Laser Beam.

Wavelength, λ (μm)	Exposure Duration, t (s)	Maximum Permissible Exposure (MPE)	Notes for Calculation and Measurement
Ultraviolet			
0.200–0.302	10^{-2}–3×10^4	3×10^{-3} J $\cdot$ cm^{-2}	
0.303	10^{-2}–3×10^4	4×10^{-3} J $\cdot$ cm^{-2}	
0.304	10^{-2}–3×10^4	6×10^{-3} J $\cdot$ cm^{-2}	
0.305	10^{-2}–3×10^4	1.0×10^{-2} J $\cdot$ cm^{-2}	1-mm limiting aperture
0.306	10^{-2}–3×10^4	1.6×10^{-2} J $\cdot$ cm^{-2}	In no case shall the total irradiance, over all the wavelengths within the UV spectral region, be greater than 1 watt per square centimeter upon the cornea
0.307	10^{-2}–3×10^4	2.5×10^{-2} J $\cdot$ cm^{-2}	
0.308	10^{-2}–3×10^4	4.0×10^{-2} J $\cdot$ cm^{-2}	
0.309	10^{-2}–3×10^4	6.3×10^{-2} J $\cdot$ cm^{-2}	
0.310	10^{-2}–3×10^4	1.0×10^{-1} J $\cdot$ cm^{-2}	
0.311	10^{-2}–3×10^4	1.6×10^{-1} J $\cdot$ cm^{-2}	
0.312	10^{-2}–3×10^4	2.5×10^{-1} J $\cdot$ cm^{-2}	
0.313	10^{-2}–3×10^4	4.0×10^{-1} J $\cdot$ cm^{-2}	
0.314	10^{-2}–3×10^4	6.3×10^{-1} J $\cdot$ cm^{-2}	
0.315–0.400	10^{-9}–10	$0.56 \, t^{1/4}$ J $\cdot$ cm^{-2}	
0.315–0.400	10–10^3	1 J $\cdot$ cm^{-2}	
0.315–0.400	10^3–3×10^4	1×10^{-3} W $\cdot$ cm^{-2}	
Visible and Near-Infrared			
0.4–1.4	10^{-9}–10^{-7}	2×10^{-2} J $\cdot$ cm^{-2}	1-mm limiting aperture
	10^{-7}–10	$1.1 \, t^{1/4}$ J $\cdot$ cm^{-2}	
	10–3×10^4	0.2 W $\cdot$ cm^{-2}	
Far-Infrared			
1.4–10^3	10^{-9}–10^{-7}	10^{-2} J $\cdot$ cm^{-2}	1-mm limiting aperture for 1.4 to 100 μm
	10^{-7}–10	$0.56 \, t^{1/4}$ J $\cdot$ cm^{-2}	11-mm limiting aperture for 0.1 to 1 mm
	>10	0.1 W $\cdot$ cm^{-2}	

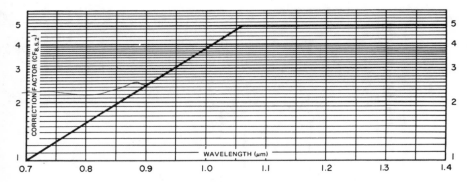

NOTE: $CF_{8.5.2} = C_A$ (see Tables 6-1 and 6-2).

Figure 6-9. Correction Factors for Wavelengths 0.7–1.4 μm ($CF_{8.5.2}$). Source: American National Standards Institute, Reference 1.

DETERMINATION OF MPE FOR REPETITIVELY PULSED LASERS

In order to determine the MPE applicable for an exposure to a repetitively pulsed laser[1,2] is is necessary that the value of the following parameters be known:

- Wavelength
- PRF
- Duration of a single pulse
- Duration of a complete exposure

The solution process necessitates performing two analyses, defined in Steps 1 and 2 that follow, and reaching a conclusion based on Step 3.

Step 1—Individual Pulse Limitation. This first analysis necessitates calculating the MPE based on the limitation that a single pulse exposure not be greater than the single pulse MPE (see Items 4a and 4e of Table 6-5) for pulses greater than 10^{-5} second; and cannot exceed this single pulse MPE multiplied by the correction factor in Figure 6-13 for pulses equal to or less than 10^{-5} second (see Item 4d of Table 6-5).

Table 6-5. Summary of MPE Correction Factors and Special Handling Techniques For Visible and Near Infrared.

Item No.	Correction Item Title	Wavelength Range	Description of Correction or Handling Technique	Applicable Table and Figures	Comments
1	When Limiting Aperture (Pupil) is not 7 mm	400–1,400 nm	No corrections made to intrabeam or extended source MPE values.	Table 6-1 and 6-2.	Represents most conservative approach.
2	Wavelength Corrections	a) All	a) No wavelength corrections are to be made to the intrabeam, extended source or skin MPE values of Tables 6-1, 6-2 and 6-4, respectively, except as shown below.	Tables 6-1, 6-2 and 6-4	
		b) 0.55 μm to 0.70 μm	b) Increase MPEs by appropriate correction factor (C_B) of Figure 6-10.	Figure 6-10; Tables 6-1 & 6-2	
		c) 0.70 μm to 1.06 μm	c) Increase MPEs by appropriate correction factor ($CF_{8.5.2} = C_A$) of Figure 6-9.	Figure 6-9, Tables 6-1 & 6-2	
		d) >1.06 μm to <1.4 μm	d) Increase MPEs by a factor of 5 (over value at 0.7 μm). See Figure 6-9.	Figure 6-9, Tables 6-1 & 6-2	
3	Exposure Duration Corrrections	a) 1.0 μm to 1.4 μm	a) For exposure times of 10^{-9} to 10^{-4} second, increase the intrabeam MPE values by a factor of 2.		
		b) 0.4 μm to 1.4 μm	b) For exposure durations between 1 and 3×10^4 seconds, the intrabeam MPE irradiance values and extended source MPE radiance values are as given in Figures 6-11 and 6-12 respectively, with all correction factors incorporated.	Figure 6-11 & 6-12	
4	Multiple Pulse Trains, Pulsed and Scanning Lasers with Multiple Exposures and PRF Greater than 1 Hertz (for Pulses within the Train)	All	The MPE for energy* or power* in multiple pulse or multiple exposure trains has the limits as specified below: a) MPE for Individual Pulse in a Train The energy* or power* is limited to the MPE for a comparable pulse, as is determined in earlier sections of this chapter (e.g., intrabeam, extended source and skin MPEs).		*Specified by irradiance (E, in W · cm^{-2}) or radiant exposure (H, in J · cm^{-2}) for intrabeam, extended source or skin exposures; and may also be specified by radiance (L, in

b) *MPE for Entire Pulse Train*

The average energy* or power* is limited to the MPE of one pulse whose duration is the same as the pulse train.

c) *MPE for Groups of Pulses Within the Train*

Limited to the MPE of one pulse the same duration as the group(s) of pulses within the train. (Same approach as item b above.)

d) *MPE for Individual Pulses With Duration Less than 10 μsec*

The MPE of the individual pulse obtained in item a above (as determined in earlier sections of this chapter on intrabeam, extended source and skin MPEs) is reduced as shown in Figure 6-13.

e) *MPE for Individual Pulse Duration 10 μsec or Greater*

The MPE for an individual pulse in a train is calculated from the MPE for the total "on time" pulse (TOTP), which has a duration equal to the sum of all the individual pulse durations in the train, as follows:

- The MPE radiant exposure or integrated exposure of an individual pulse within the train is reduced to the MPE for the TOTP divided by the number of pulses within the train.

- A further limitation is that the average irradiance in the pulse train cannot exceed the MPE as defined in item b above, and the MPEs for the individual pulses must be reduced to keep within its limitation.

Figure 6-13

5 Repetitive Pulses at PRF of Less than 1 Hertz   All   MPE should be considered additive over a 24-hour period.

W · cm^{-2} · sr^{-1}) or integrated radiance (L_p, in J · cm^{-2} · sr^{-1}) for extended source exposures.

Note: (1) Compiled from information contained in Reference 1.

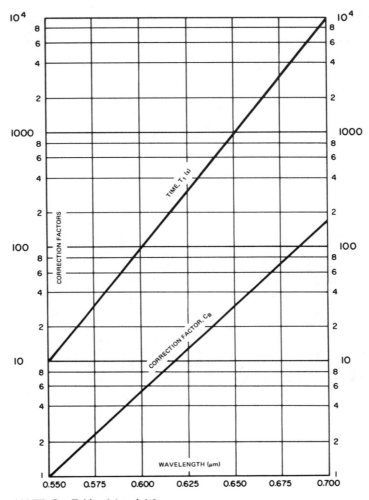

NOTE: See Tables 6-1 and 6-2.

Figure 6-10. Correction factors for wavelengths 0.55–0.7 μm. Source: American National Standards Institute, Reference 1.

Step 2—Average Irradiance or Average Radiant Exposure Limitation. The average power limitation necessitates calculating the average irradiance or total radiant exposure for the entire pulse train for comparison with the MPE applicable for the duration of the entire exposure (see Item 4e of Table 6-5).

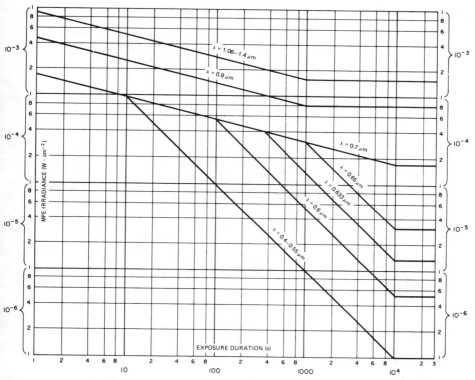

Figure 6-11. Ocular MPE for intrabeam viewing as a function of exposure duration and wavelengths. Source: American National Standards Institute, Reference 1.

Step 3—Conclusion. Compare the analyses of Step 1 and Step 2. The limitation that provides the lowest total exposure is then used, and provides the most conservative approach.

MPE CORRECTION FACTORS—INFRARED

Few data exist in the infrared range of 1,400 nm to 1 mm for making MPE corrections. At 1,540 nm the MPEs given in Tables 6-1 and 6-2 for intrabeam viewing and extended source viewing, respectively, are increased by a factor of 100 for time periods shorter than 1 μsec. However, on the basis of existing information no extrapolation to other wavelengths is justified.

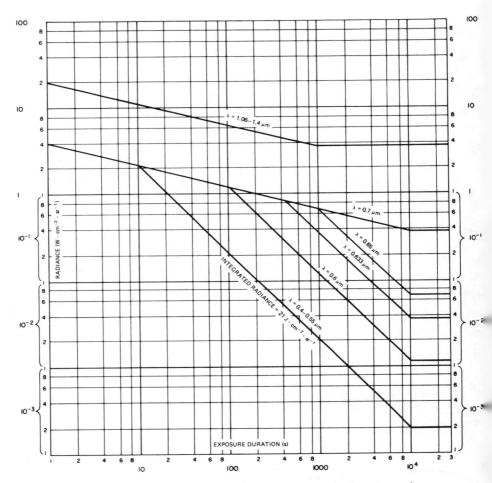

Figure 6-12. Ocular MPE for extended sources as a function of exposure duration and wavelengths. National Standards Institute, Reference 1.

Example 6-4. Single Pulse Near Infrared Laser. Find the intrabeam direct viewing MPE for a 1,064-nm N_d :YAG laser with a pulse duration of 8×10^{-4} second.

1) Since the pulse duration of 8×10^{-4} second is the same as that for the ruby laser of Example 6-1, the MPE for a single pulse differs only by the wavelength correction factor.

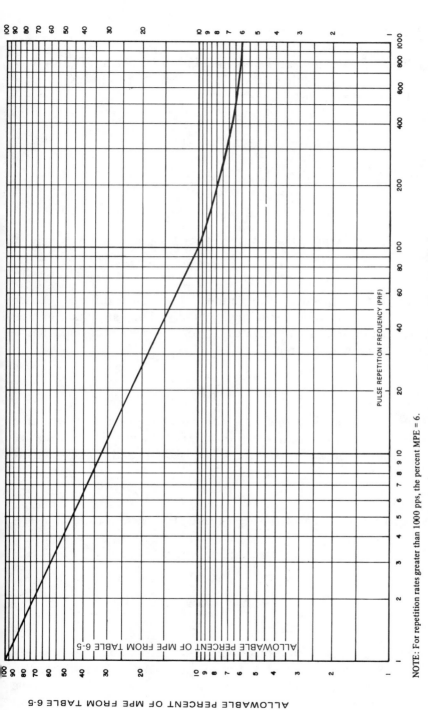

Figure 6-13. Reduction of MPE for repetitively pulsed or multiple exposures from scanning lasers, individual pulse or exposure less than 10 μsec. Source: American National Standards Institute, Reference 1.

NOTE: For repetition rates greater than 1000 pps, the percent MPE = 6.

PULSE REPETITION FREQUENCY (PRF)

ALLOWABLE PERCENT OF MPE FROM TABLE 6-5

75

2) From Figure 6-9, at 1,064 nm the direct beam MPE for this laser is five times that for the visible laser of Example 6-1.

3) From Example 6-1,

$$\text{MPE: } H = 8.56 \times 10^{-6} \text{J} \cdot \text{cm}^{-2}$$

$$\text{MPE: } E = 1.07 \times 10^{-2} \text{W} \cdot \text{cm}^{-2}$$

4) Thus, for N_d : YAG laser,

$$\text{MPE: } H = 5(8.56 \times 10^{-6} \text{J} \cdot \text{cm}^{-2})$$

$$= 4.28 \times 10^{-5} \text{J} \cdot \text{cm}^{-2}$$

$$\text{MPE: } E = 5(1.07 \times 10^{-2} \text{W} \cdot \text{cm}^{-2})$$

$$= 5.35 \times 10^{-2} \text{W} \cdot \text{cm}^{-2}$$

Example 6-5. Spectral Corrections For Near-Infrared Laser MPEs. Find the MPE for a single pulse gallium arsenide laser of 200 nsec duration, that operates at room temperature and has a peak wavelength at 904 nm.

1) $$\lambda = 904 \text{ nm}$$

$$t = 200 \text{ nsec} = 0.2 \times 10^{-6} \text{ second}$$

2) From Table 6-1, the MPE for direct viewing of the source for exposure duration of 0.2×10^{-6} second (i.e., exposure duration between 10^{-9} and 2×10^{-5} second) is the radiant exposure

$$H = 5C_A \times 10^{-7} \text{J} \cdot \text{cm}^{-2}$$

3) From Figure 6-9, for $\lambda = 904$ nm, the spectral correction factor is 2.5.

4) Thus, the MPE for intrabeam viewing is

$$\text{MPE: } H = (2.5)(5 \times 10^{-7} \text{J} \cdot \text{cm}^{-2})$$

$$= 1.25 \times 10^{-6} \text{J} \cdot \text{cm}^{-2}$$

5) Extended Source MPE

- For an exposure duration of 0.2×10^{-6} second, the limiting angular subtense as determined from Figure 6-4 is 3.3 milliradians.
- Extended source MPE applies for sources subtending an angle greater than 3.3 milliradians.

- From Figure 6-8, the MPE for extended source viewing for exposure duration of 0.2×10^{-6} second (2×10^{-7} second) gives an integrand radiance of $6 \times 10^{-2} J \cdot cm^{-2} \cdot sr^{-1}$.
- Similar to the case of intrabeam viewing, the spectral correction factor of 2.5 of step (3) must be applied.
- Thus, the MPE for extended source viewing is

$$\text{MPE: } L_p = (2.5)(6 \times 10^{-2} J \cdot cm^{-2} \cdot sr^{-1})$$
$$= 1.5 \times 10^{-1} J \cdot cm^{-2} \cdot sr^{-1}$$

Example 6-6. Repetitively Pulsed Visible Laser With Very High PRF. Find the direct intrabeam MPE of a 514.5 nm argon laser for a 0.25-second total exposure, PRF of 10 MHz and pulse duration of 10 nsec. From above:

$\lambda = 514.5$ nm
$T = 0.25$ second (total exposure duration of pulse train)
$F = \text{PRF} = 10 \, \text{MHz} = 10^7 \, \text{Hz}$
$t = 10 \, \text{nsec} = 10^{-8}$ second

Step 1—Individual Pulse Limitation

- From Table 6-1 or Figure 6-5, for $\lambda = 514.5$ nm (i.e., between 400 and 700 nm), and a single pulse exposure of 10^{-8} second (i.e., between 10^{-9} and 1.8×10^{-5} second),

$$\text{MPE/Pulse: } H = 5 \times 10^{-7} J \cdot cm^{-2}$$

- Since the individual pulse is less than 10 μsec, a reduced single pulse MPE results and is obtained from Figure 6-13. From Figure 6-13, for a PRF $\geq$ 1000 Hz (e.g., 10^7 Hz) the MPE is reduced to 6% of the above value. Thus,

$$\text{MPE/Pulse: } H_1 = (0.06)(5 \times 10^{-7} J \cdot cm^{-2})$$
$$= 3 \times 10^{-8} J \cdot cm^{-2}$$

- The MPE for the total 0.25-second exposure is

$$\text{MPE/Train: } H = H_1 \cdot F \cdot T$$
$$= (3 \times 10^{-8} J \cdot cm^{-2})(10^7 \, \text{Hz})(0.25 \, \text{second})$$
$$= 7.5 \times 10^{-2} J \cdot cm^{-2}$$

Step 2—Average Radiant Exposure Limitation

- The average radiant exposure for the 0.25-second pulse train is limited to the MPE for one pulse whose duration is the same as the 0.25-second pulse train.
- From Table 6-1, for a 0.25-second exposure duration (i.e., between 1.8×10^{-5} and 10 seconds)

$$\text{MPE (average): } H = 1.8\,t^{3/4} \times 10^{-3}\,\text{J} \cdot \text{cm}^{-2}$$
$$= 1.8(0.25)^{3/4} \times 10^{-3}\,\text{J} \cdot \text{cm}^{-2}$$
$$= 1.8(0.35) \times 10^{-3}\,\text{J} \cdot \text{cm}^{-2}$$
$$= 6.3 \times 10^{-4}\,\text{J} \cdot \text{cm}^{-2}$$

Use of Figure 6-5 yields the same value.

Step 3—Conclusion

- Since Step 2 gives the more restrictive or limiting case, the correct MPE is

$$\text{MPE: } H = 6.3 \times 10^{-4}\,\text{J} \cdot \text{cm}^{-2} \text{ in 0.25 second}$$

or

$$E = \frac{H}{T} = \frac{6.3 \times 10^{-4}\,\text{J} \cdot \text{cm}^{-2}}{0.25 \text{ second}}$$
$$= 2.52 \times 10^{-3}\,\text{W} \cdot \text{cm}^{-2}$$

Example 6-7. Repetitively Pulsed Near-Infrared Laser With Moderate PRF. Find the intrabeam direct viewing MPE for a 905-nm Ga As laser with a PRF of 1 kHz and pulse duration of 100 nsec. From above:

$\lambda = 905$ nm
$F = \text{PRF} = 1 \text{ kHz} = 10^3 \text{ Hz}$
$t = 100 \text{ nsec} = 0.1 \times 10^{-6}$ second

Step 1—Individual Pulse Limitation

- From Table 6-1, for $\lambda = 905$ nm and pulse duration of 0.1×10^{-6} (i.e., between 10^{-9} and 1.8×10^{-5} second),

$$\text{MPE/Pulse: } H = 5\,C_A \times 10^{-7}\,\text{J} \cdot \text{cm}^{-2}$$

- From Figure 6-9, Correction Factor A (C_A) is 2.5 at $\lambda = 905$ nm. Thus,

$$\text{MPE/Pulse:} \quad H = (5)(2.5)(10^{-7}\,\text{J} \cdot \text{cm}^{-2})$$
$$= 1.25 \times 10^{-6}\,\text{J} \cdot \text{cm}^{-2}$$

- Since $t < 10$ μsec, the single pulse MPE reduction factor is obtained from Figure 6-13. For PRF = 1,000 Hz, the reduction factor is 0.06. Thus, the MPE for an individual pulse is

$$\text{MPE/Pulse:} \quad H_1 = (0.06)(1.25 \times 10^{-6}\,\text{J} \cdot \text{cm}^{-2})$$
$$= 7.5 \times 10^{-8}\,\text{J} \cdot \text{cm}^{-2}$$

- Since the 905-nm wavelength is out of the visible region, a natural aversion response is not provided as it would be with a visible wavelength laser. Assume a 10-second exposure duration for this particular laser application, or

$$T = 10 \text{ seconds}$$

- The MPE for a 10-second pulse train exposure is

$$\text{MPE/Train:} \quad H = H_1 \cdot F \cdot T$$
$$= (7.5 \times 10^{-8}\,\text{J} \cdot \text{cm}^{-2})(10^3 \text{ Hz})\,(10 \text{ seconds})$$
$$= 7.5 \times 10^{-4}\,\text{J} \cdot \text{cm}^{-2}$$

Step 2—Average Power Limitation

- From Table 6-1, for a 10-second pulse train exposure (i.e., between 1.8×10^{-5} and 10^3 seconds) at $\lambda = 905$ nm,

$$\text{MPE (average):} \quad H = 1.8\,C_A\,t^{3/4} \times 10^{-3}\,\text{J} \cdot \text{cm}^{-2}$$
$$= (1.8)(2.5)(10)^{3/4}(10^{-3}\,\text{J} \cdot \text{cm}^{-2})$$
$$= (4.5)(5.6)(10^{-3}\,\text{J} \cdot \text{cm}^{-2})$$
$$= 2.5 \times 10^{-2}\,\text{J} \cdot \text{cm}^{-2}$$

- or

$$E = \frac{H}{T} = \frac{2.5 \times 10^{-2}\,\text{J} \cdot \text{cm}^{-2}}{10 \text{ seconds}}$$
$$= 2.5 \times 10^{-3}\,\text{W} \cdot \text{cm}^{-2}$$

Step 3—Conclusion

- Evidently the limitation of Step 1 determines the MPE.

$$\text{MPE/Train: } H = 7.5 \times 10^{-4} \, \text{J} \cdot \text{cm}^{-2}$$

or

$$E = \frac{H}{T} = \frac{7.5 \times 10^{-4} \, \text{J} \cdot \text{cm}^{-2}}{10 \text{ seconds}}$$

$$= 7.5 \times 10^{-5} \, \text{W} \cdot \text{cm}^{-2}$$

Example 6-8. Low PRF, Long Pulse, Repetitively Pulsed Visible Laser.   Find the MPE for a 632.8-nm He-Ne laser with a PRF of 100 Hz, pulse duration of 10^{-3} second and total exposure duration of the pulse train of 0.25 second. From above:

$\lambda = 632.8$ nm
$F = 100$ Hz
$t = 10^{-3}$ second
$T = 0.25$ second

Step 1—Individual Pulse Limitation

- Since the individual pulse duration $\geq$ 10 μsec, then the total "on time" T_i for the 0.25-second exposure must be obtained.

$$T_i = t \cdot F \cdot T \qquad \qquad \text{(Eq. 6-6)}$$

$$= (10^{-3} \text{ second})(100 \text{ Hz})(0.25 \text{ second})$$

$$= 2.5 \times 10^{-2} \text{ second}$$

- For exposure time $T_i = 2.5 \times 10^{-2}$ second and $\lambda = 632.8$ (i.e., between 400 and 700 nm), from Figure 6-5,

$$\text{MPE } (T_i): \quad H = 10^{-4} \, \text{J} \cdot \text{cm}^{-2}$$

$$\text{MPE/Train: } H = 10^{-4} \, \text{J} \cdot \text{cm}^{-2}$$

- With a PRF = 100 Hz and a 0.25-second exposure duration, 25 pulses occur. The MPE for a single pulse is then

$$\text{MPE/Pulse:} \quad H_i = \frac{\text{MPE/Train}}{\text{No. of pulses in exposure duration}}$$

$$= \frac{10^{-4}\,\text{J} \cdot \text{cm}^{-2}}{25 \text{ pulses}}$$

$$= 4 \times 10^{-6}\,\text{J} \cdot \text{cm}^{-2}$$

- For a single 10^{-3}-second pulse not in a train, from Figure 6-5,

$$\text{MPE/Pulse:} \quad H = 10^{-5}\,\text{J} \cdot \text{cm}^{-2}$$

Thus, the MPE has been reduced by a factor of 2.5.

Step 2—Average Power Limitation

For a 0.25-second exposure the MPE from Figure 6-5 is,

$$\text{MPE:} \quad H = 6.3 \times 10^{-4}\,\text{J} \cdot \text{cm}^{-2}$$

Step 3—Conclusion

- Step 1 defines the MPE, as this is the more restrictive case. Thus,

$$\text{MPE}\,(T_i)\text{:} \quad H = 10^{-4}\,\text{J} \cdot \text{cm}^{-2}$$

- MPE (average power): $\quad E_{\text{avg}} = H_i \cdot F \qquad\qquad$ (Eq. 6-7)

$$= (4 \times 10^{-6}\,\text{J} \cdot \text{cm}^{-2})(100 \text{ Hz})$$

$$= 4 \times 10^{-4}\,\text{W} \cdot \text{cm}^{-2}$$

Example 6-9. One Pulse Group, Short Pulse Laser. Determine the MPE of a Q-switched ruby laser of wavelength 694.3 nm that outputs three 20-nsec pulses each separated by 100 nsec. This is

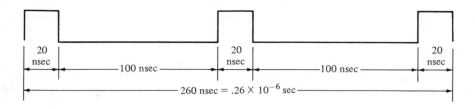

not a repetitively pulsed laser in the usual sense, for example, one having a continuous train of pulses lasting about 0.25 second or more with the pulses being reasonably spaced. This group of pulses should be considered as one pulse (i.e., $t = .26 \times 10^{-6}$ second) and the MPE determination handled as follows:

- From Table 6-1, the intrabeam visible MPE for an exposure between 10^{-9} and 1.8×10^{-5} second (e.g., $.26 \times 10^{-6}$ second) for λ between 400 and 700 nm is

$$\text{MPE: } H = 5 \times 10^{-7} \text{J} \cdot \text{cm}^{-2}$$

- The MPE per pulse is

$$\text{MPE/Pulse: } H = \frac{5 \times 10^{-7} \text{J} \cdot \text{cm}^{-2}}{3 \text{ pulses}} = 1.7 \times 10^{-7} \text{J} \cdot \text{cm}^{-2}$$

Example 6-10. Repetitively Pulsed Pulse Groups.   Determine the MPE of a 488-nm argon laser used in a pulse code modulated (pcm) communication link. The laser presents 10^4 "words" per second (for example, 10^4 pulse groups per second), and each word consists of five 20-nsec pulses spaced at coded intervals so that each pulse group lasts no longer than 1 μsec.

Step 1—Individual Pulse (Group) Limitation

- Using the approach of Example 6-9, each pulse group or word will be considered a single pulse ($t = 1$ μsec $= 10^{-6}$ second). Thus, the PRF $= 10^4$ Hz or words per second. Also, from Table 6-1, the intrabeam visible MPE for an exposure between 10^{-9}

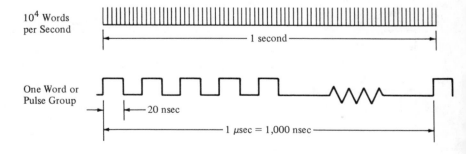

and 1.8×10^{-5} (e.g., 10^{-6} second), for λ between 400 and 700 nm is

$$\text{MPE: } H = 5 \times 10^{-7} \text{J} \cdot \text{cm}^{-2}$$

- Since the individual pulse group (word) is less than 10 μsec (i.e., $t = 1$ μsec), a reduced single pulse MPE results and is obtained from Figure 6-13. From that figure, for a PRF of 10^4 Hz (i.e., $\geq 1,000$ Hz), the MPE is reduced to 6% of the above value, or

$$\text{MPE (pulse group): } H_i = (0.06)(5 \times 10^{-7} \text{J} \cdot \text{cm}^{-2})$$

$$= 3 \times 10^{-8} \text{J} \cdot \text{cm}^{-2}$$

- From Eq. 6-7,

$$\text{MPE (average power): } E_{\text{avg}} = H_i \cdot F$$

$$= (3 \times 10^{-8} \text{J} \cdot \text{cm}^{-2})(10^4 \text{ Hz})$$

$$= 3 \times 10^{-4} \text{W} \cdot \text{cm}^{-2}$$

Step 2—Average Power Limitation

- For a 0.25-second exposure the MPE from Figure 6-5 is

$$\text{MPE: } H = 6.3 \times 10^{-4} \text{J} \cdot \text{cm}^{-2}$$

- $$\text{MPE: } E = \frac{H}{T} = \frac{6.3 \times 10^{-4} \text{J} \cdot \text{cm}^{-2}}{0.25 \text{ second}}$$

$$= 2.52 \times 10^{-3} \text{W} \cdot \text{cm}^{-2}$$

Step 3—Conclusion

Since Step 1 gives the more restrictive case, it is applied, and

$$\text{MPE (average power)} = 3 \times 10^{-4} \text{W} \cdot \text{cm}^{-2}$$

FORMULAS, CONSIDERATIONS AND EXAMPLES USEFUL IN EVALUATION OF VARIOUS LASER APPLICATIONS*

The following areas will be addressed, in which many useful and practical considerations and formulas will be discussed and developed, which are helpful in the context of laser safety assessment.

*Adapted from Reference 2.

- Atmospheric attenuation of laser beam
- Average irradiance and radiant exposure as a function of range
- Reflected irradiance or radiant exposure for a diffuse target
- Optical system viewing of laser radiation
- Single exposure from a scanning laser

Where deemed particularly useful, problems are provided.

Atmospheric Attenuation of Laser Beam

The radiant exposure (H) or the beam irradiance (E) for a nondivergent beam at range r that is attenuated by the atmosphere is given by

$$H = H_0 e^{-\mu r} \qquad \text{(Eq. 6-8)}$$

or

$$E = E_0 e^{-\mu r} \qquad \text{(Eq. 6-9)}$$

where

H_0 = emergent beam radiant exposure at zero range measured in $J \cdot cm^{-2}$ for pulsed lasers

E_0 = emergent beam irradiance at zero range measured in $W \cdot cm^{-2}$ for CW lasers

μ = atmospheric attenuation coefficient (cm^{-1}) at a particular wavelength

r = range from the laser to the viewer or to a diffuse target (cm)

The attenuation coefficient, u, varies from 10^{-4} for thick fog to 10^{-7} in air with very good visibility.

Average Irradiance and Radiant Exposure As a Function of Range

The average irradiance at range r for a direct circular beam is the total power in the beam at that range divided by the area of the beam at that range, and is given by

$$E = \frac{\Phi e^{-\mu r}}{\pi \left[\dfrac{a + r\phi}{2}\right]^2} = \frac{1.27\Phi e^{-\mu r}}{(a + r\phi)^2} \qquad \text{(Eq. 6-10)}$$

Similarly, the radiant exposure in a nonturbulent medium is the total energy in the beam at that range divided by the total area, and is given by

$$H = \frac{Qe^{-\mu r}}{\pi \left[\dfrac{a + r\phi}{2} \right]^2} = \frac{1.27Qe^{-\mu r}}{(a + r\phi)^2} \qquad \text{(Eq. 6-11)}$$

where

E = irradiance at range r, measured in $J \cdot cm^{-2}$ for pulsed lasers and $W \cdot cm^{-2}$ for CW lasers

Φ = total radiant power (or radiant flux) output of a CW laser, or average radiant power of a repetitively pulsed laser, measured in watts

H = radiant exposure at range r, measured in $J \cdot cm^{-2}$ for pulsed lasers and $W \cdot cm^{-2}$ for CW lasers

Q = total radiant energy output of a pulsed laser, measured in joules

e = base of natural logarithms

r = range from the laser to the viewer or to a diffuse target in cm

a = diameter of emergent laser beam in cm

ϕ = emergent beam divergence measured in radians

μ = atmospheric attenuation coefficient (cm^{-1}) at a particular wavelength

Equations 6-10 and 6-11 are accurate only for small emergent beam divergence (ϕ) values; that is, accuracy better than 1% for angles below 0.17 radian (10°) and better than 5% for angles less than 0.37 radian (21°).

Example 6-11.   Determine the radiant exposure at 1 km of a 0.1-joule ruby laser with a beam divergence of 1 milliradian and an emergent beam diameter of 0.7 cm. Assume very clear visibility so that $\mu = 10^{-7}$ cm^{-1}.

From above:

$r = 1$ km $= 10^5$ cm

$Q = 0.1$ joule

$\phi = 1$ milliradian $= 10^{-3}$ radian

$a = 0.7$ cm

$\mu = 10^{-7}$ cm^{-1}

Using Eq. 6-11,

$$H = \frac{1.27 Q e^{-\mu r}}{(a + r\phi)^2}$$

$$= \frac{1.27(0.1) e^{-(10^{-7})(10^{-5})}}{[0.7 + (10^5)(10^{-3})]^2}$$

$$= 1.25 \times 10^{-5} \, \text{J} \cdot \text{cm}^{-2}$$

Reflected Irradiance or Radiant Exposure for a Diffuse Target

The reflected irradiance or radiant exposure for a diffuse reflector, where $r_1 \gg D_L$ (see Figure 6-3) is given by

$$E = \frac{\Phi \rho_\lambda \cos \theta_v}{\pi r_1^2} \qquad \text{(Eq. 6-12)}$$

and

$$H = \frac{Q \rho_\lambda \cos \theta_v}{\pi r_1^2} \qquad \text{(Eq. 6-13)}$$

Example 6-12. Determine the maximum reflected radiant exposure from a diffuse target of reflectance 0.6, placed a distance of 10 meters from the operator. The target is normal to the line of fire (i.e., $\theta_v = 0°$), and the energy striking the target is 0.1 joule.
From above:

$\rho_\lambda = 0.6$
$r_1 = 10$ meters $= 10^3$ cm
$Q = 0.1$ joule
$\cos \theta_v = \cos (0°) = 1$

From Eq. 6-13:

$$H = \frac{Q \rho_\lambda \cos \theta_v}{\pi r_1^2}$$

$$= \frac{(0.1 \, \text{J})(0.6)(1)}{(3.14)(10^{-3} \, \text{cm})^2}$$

$$= 1.91 \times 10^{-8} \, \text{J} \cdot \text{cm}^{-2}$$

Optical System Viewing of Laser Radiation

The ratio, G, of the radiant exposure or irradiance at the retina when viewing is aided by an optical instrument or system to that received if the eye is unaided, is defined below.

For intrabeam viewing and specular reflections (or for a diffuse spot not resolved by the eye and optical system), the ratio G is given by:

$$G = \frac{D_o^2}{d_e^2} \quad \text{for} \quad d_e \geq D_e \qquad \text{(Eq. 6-14)}$$

and

$$G = \frac{D_o^2}{D_e^2} = P^2 \quad \text{for} \quad d_e \leq D_e \qquad \text{(Eq. 6-15)}$$

where

D_o = diameter of objective of an optical system (cm)
d_e = diameter of the pupil of the eye (varies from approximately 0.2 to 0.7 cm)
D_e = diameter of the exit pupil of an optical system (cm)
P = magnifying power of an optical system or instrument

For indirect viewing of a diffuse reflection, extended objects only, that is, the object subtends an angle greater than 0.6 milliradian when magnified, the ratio G is given by,

$$G = \frac{D_o^2}{P^2 d_e^2} \leq 1 \quad \text{for} \quad d_e \geq D_e \qquad \text{(Eq. 6-16)}$$

and

$$G = \frac{D_o^2}{P^2 D_e^2} = 1 \quad \text{for} \quad d_e \leq D_e \qquad \text{(Eq. 6-17)}$$

Note that the ratio G is influenced by the optical transmission of the instrument, but this factor is usually insignificant. If this is not the case, then this factor must be in the numerators of Eqs. 6-14 through 6-17.

Example 6-13.   For the laser of Example 6-12, find the relative hazard to the viewer's eyes, during night viewing, for the diffuse reflection, when a pair of 10×50 binoculars (that is, $P = 10$ and $D_o = 50$ mm) is used.

1) For night viewing the pupil diameter is at maximum opening so that $d_e = 0.7$ cm.
2) From Eq. 6-15, the exit pupil of the optical instrument is

$$D_e = \frac{D_o}{P} = \frac{5.0 \text{ cm}}{10} = 0.5 \text{ cm}$$

3) Since $d_e \geqq D_e$, use Eq. 6-16.

$$G = \frac{D_o^2}{P^2 d_e^2} = \frac{(5.0 \text{ cm})^2}{(10)^2 (0.7 \text{ cm})^2} = 0.51$$

4) From Example 6-12, a radiant expose of $H = 1.91 \times 10^{-8}$ J $\cdot$ cm^{-2} was available at a return distance of 10 meters from the target.
5) Thus, using the binoculars, the hazard is equivalent to a corneal radiant exposure to the naked eye of

$$H = (0.51)(1.91 \times 10^{-8} \text{ J} \cdot \text{cm}^{-2})$$
$$= .98 \times 10^{-8} \text{ J} \cdot \text{cm}^{-2}$$

Example 6-14.   If an operator were to view a specularly reflected beam through a pair of 7×50 binoculars, at a point where the beam radiant exposure measures 2×10^{-9} J $\cdot$ cm^{-2}, find the relative hazard compared to unaided viewing. From above:

$P = 7$
$D_o = 50$ mm $= 5$ cm
$H = 2 \times 10^{-9}$ J $\cdot$ cm^{-2}

1) From Eq. 6-15,

$$D_e = \frac{D_o}{P} = \frac{5 \text{ cm}}{7} = 0.71 \text{ cm}$$

2) For day viewing ($d_e \approx 0.2$ cm) or night viewing ($d_e \approx 0.7$ cm), $d_e \leq D_e$. Therefore, use Eq. 6-15,

$$G = P^2 = (7)^2 = 49$$

3) Thus, the operator would be viewing a level 49 times greater than with the naked eye or a corneal radiant exposure of

$$H = 49(2 \times 10^{-9} \text{ J} \cdot \text{cm}^{-2})$$
$$= .98 \times 10^{-7} \text{ J} \cdot \text{cm}^{-2}$$

Single Exposure From A Scanning Laser

The corneal radiant exposure for a single exposure from a scanning laser beam is given by

$$H = \frac{1.27\Phi e^{-\mu r}}{(a + r\phi)^2} \cdot \frac{d_e}{rS\theta_s} \quad \text{for} \quad d_e > (a + r\phi) \quad \text{(Eq. 6-18)}$$

or

$$H = \frac{1.27\Phi e^{-\mu r}}{(a + r\phi)(rS\theta_s)} \quad \text{for} \quad d_e < (a + r\phi) \quad \text{(Eq. 6-19)}$$

where

S = scan rate of a scanning laser (number of scans across eye per second)

θ_s = maximum angular sweep of a scanning beam (radians)

Other parameters are defined on pages 51 to 53.

The applicable MPEs are dependent on the repetitive nature of the exposure and the exposure duration of a single pulse, where

$$T = \frac{(a + r\phi)}{rS\theta_s} \quad \text{for} \quad d_e \leq (a + r\phi) \quad \text{(Eq. 6-20)}$$

or

$$T = \frac{d_e}{rS\theta_s} \quad \text{for} \quad d_e \geq (a + r\phi) \quad \text{(Eq. 6-21)}$$

and the PRF is S if each scan passes over the eye.

Example 6-15. Determine the exposure of a scanning He-Ne laser having the following parameters:

$a = 0.1$ cm
$\phi = 5 \times 10^{-3}$ radian

$\Phi = 5 \text{ mW} = 5 \times 10^{-3} \text{ W}$
$\theta_s = 0.1 \text{ radian}$
$S = 30 \text{ second}^{-1} = 30 \text{ Hz}$
Intrabeam viewing distance: $r = 200$ cm

1) From Eq. 6-3, the beam diameter is

$$D_L = (a + r\phi)$$

$$= 0.1 \text{ cm} + (200 \text{ cm})(5 \times 10^{-3} \text{ radian})$$

$$= 1.1 \text{ cm}$$

2) Since d_e varies from about 0.2 to 0.7 cm,

$$d_e \le (a + r\phi)$$

and Eqs. 6-19 and 6-20 must be used.

3) The PRF at the eye is 30 pulses per second (i.e., $S = 30$), and the exposure time of a single pulse, obtained from Eq. 6-20, is

$$T = \frac{(a + r\phi)}{rS\theta_s}$$

$$= \frac{(0.1 \text{ cm}) + (200 \text{ cm})(5 \times 10^{-3} \text{ radian})}{(200 \text{ cm})(30 \text{ sec}^{-1})(0.1 \text{ radian})} = \frac{1.1}{600}$$

$$= 1.83 \times 10^{-3} \text{ second}$$

4) The radiant exposure from Eq. 6-19 is

$$H = \frac{1.27\Phi e^{-\mu r}}{(a + r\phi)(rS\theta_s)}$$

$$= \frac{(1.27)(5 \times 10^{-3})(1)}{[0.1 + (200)(5 \times 10^{-3})] [(200)(30)(0.1)]}$$

$$= 9.6 \times 10^{-6} \text{ J} \cdot \text{cm}^{-2}$$

where $e^{-\mu r} \approx 1$ for small r.

5) The average irradiance at the cornea is

$$E_{avg} = H \cdot S$$

$$= (9.6 \times 10^{-6} \text{ J} \cdot \text{cm}^{-2})(30 \text{ sec}^{-1})$$

$$= 2.88 \times 10^{-4} \text{ W} \cdot \text{cm}^{-2}$$

6) The applicable MPE for a 0.25-second exposure is found by the cumulative exposure of almost eight pulses for a scan rate of 30. Thus,

$$H = (8)(9.6 \times 10^{-6} \, \text{J} \cdot \text{cm}^{-2})$$
$$= 7.7 \times 10^{-5} \, \text{J} \cdot \text{cm}^{-2}$$

This radiant exposure must then be compared with the MPE for a single pulse of duration

$$\tau = (8)(1.83 \times 10^{-3} \text{ second})$$
$$= 1.46 \times 10^{-2} \text{ second}$$

From Figure 6-5, for an exposure duration of $\tau = 1.46 \times 10^{-2}$ second,

$$\text{MPE:} \quad H \approx 7.7 \times 10^{-5} \, \text{J} \cdot \text{cm}^{-2}$$

Thus, the exposure is permissible for momentary or unintentional viewing.

REFERENCES

1. "American National Standard for the Safe Use of Lasers, Z 136.1-1976," American National Standards Institute, New York, 1976.
2. "Control of Hazards to Health from Laser Radiation," U.S. Department of the Army Technical Bulletin TB-MED-279, Washington, D.C., May 1975.

BIBLIOGRAPHY

1. *Laser Safety Guide*, Third Edition, Laser Institute of America, Cincinnati, Ohio, Jan. 1976.

7

Laser Beam Hazard Evaluation and Classification

INTRODUCTION

The overall evaluation of the laser beam hazard is governed by three key elements:

- The potential of the laser beam to injure personnel. This aspect is addressed by using a standardized classification of the relative laser beam hazard.
- The laser environment.
- The personnel present in the laser environment.

For the last two items, a standardized approach is difficult because of the numerous environments in which the laser can be used, and the many situations in which personnel can be exposed to the laser beam.

It follows that proper consideration of the above three elements greatly influences the measures used to control the laser beam hazard.

For a complete laser hazard evaluation it is necessary to consider the complete range of laser hazards, not only the beam hazard. Thus, all associated laser hazards such as electrical shock and airborne contaminants must be considered. Such aspects have been covered in Chapter 4; control measures are covered in Chapter 9.

LASER CLASSIFICATION CONSIDERATIONS

The Bureau of Radiological Health (BRH)[1] requires that laser products manufactured after August 2, 1976 be classified by the manufacturer in accordance with specified procedures, which requires a knowledge of laser output parameters. Thus, for product users of recently manufactured equipment, measurements and/or calculations for the purpose of laser beam hazard classification are not generally needed. However, such measurements and/or calculations may be required in any of the following circumstances:

- When modifications to the laser have been made and the possibility exists that the classification has changed.
- When beam divergence is uncertain.
- When manufacturer's information is lacking or is ambiguous.
- For outdoor usage at significant ranges where atmospheric propagation effects raise questions as to the radiation levels. See Chapter 15, which treats atmospheric effects on laser beams.
- When the laser power or energy output nears the dividing line between laser classifications, unless hazard control measures are taken for the higher-class laser.
- To verify low leakage levels in "enclosed" or stowed test positions for scanning lasers with normally hazardous radiation levels.
- When uncertainties in laser target reflectivity, dispersion characteristics of the reflected beam and other considerations require measurements for assessing hazardous radiation levels.

When measurements are necessary for the purpose of assessing the laser hazard classification, they must be made by personnel with suitable background and experience, and with the controls set to provide the maximum normal laser output power or energy levels. Further, the hazard evaluation must take into account the magnitude of the estimated error in measuring the radiation level. This estimated error would then be added to the measured radiation level for hazard evaluation purposes.

Parameters specified by the BRH[2] in reporting levels of accessible radiation for the classification of a laser product follow:

- Wavelength(s) or range, with classification based on the highest or most hazardous output level.
- For other than CW output (e.g., pulsed, Q-switched, mode locked, and so on):
  - Pulse width, with rise and decay times
  - Maximum PRF
  - Average energy per pulse, and pulse-to-pulse energy variation from average
  - PRF at which the individual pulse energy is maximum
  - PRF at which average radiant power is maximum
- Emission duration
- Measured accessible emission level
- Aperture stop diameter used for radiation measurement

For extended source lasers, the following additional parameters are required for classification:

- Laser source radiance or integrated radiance
- Maximum viewing angle subtense

LASER CLASSIFICATION DEFINITIONS

Four laser classifications have been established by the BRH in their "Laser Products Performance Standards."[1] Classification definitions given in this chapter are based on and closely follow that document, which is also the source of appropriate data tables for each class.

Class I Laser

This class laser does not provide human access to laser radiation in excess of the accessible emission limits given for Class I lasers in Table 7-1A, for any combination of emission duration and wavelength range. See Table 7-2A for wavelength and sampling interval-dependent correction factors k_1 and k_2. Table 7-2A is also applicable

Table 7-1A. Class I Accessible Emission Limits for Laser Radiation.

Wavelength (nanometers)	Emission Duration (seconds)	Class I—Accessible Emission Limits
>250 but ≤400	≤3.0 × 10⁴	$2.4 \times 10^{-5} k_1 k_2$ J*
	>3.0 × 10⁴	$8.0 \times 10^{-10} k_1 k_2 t$ J
>400 but ≤1,400	>1.0 × 10⁻⁹ to 2.0 × 10⁻⁵	$2.0 \times 10^{-7} k_1 k_2$ J
	>2.0 × 10⁻⁵ to 1.0 × 10¹	$7.0 \times 10^{-4} k_1 k_2 t^{3/4}$ J
	>1.0 × 10¹ to 1.0 × 10⁴	$3.9 \times 10^{-3} k_1 k_2$ J
	>1.0 × 10⁴	$3.9 \times 10^{-7} k_1 k_2 t$ J
	OR**	
	>1.0 × 10⁻⁹ to 1.0 × 10¹	$10 k_1 k_2 t^{1/3}$ J cm^{-2} sr^{-1}
	>1.0 × 10¹ to 1.0 × 10⁴	$20 k_1 k_2$ J cm^{-2} sr^{-1}
	>1.0 × 10⁴	$2.0 \times 10^{-3} k_1 k_2 t$ J cm^{-2} sr^{-1}
>1,400 but ≤13,000	>1.0 × 10⁻⁹ to 1.0 × 10⁻⁷	$7.9 \times 10^{-5} k_1 k_2$ J
	>1.0 × 10⁻⁷ to 1.0 × 10¹	$4.4 \times 10^{-3} k_1 k_2 t^{1/4}$ J
	>1.0 × 10¹	$7.9 \times 10^{-4} k_1 k_2 t$ J

*Class I accessible emission limits for the wavelength range of greater than 250 nm but less than or equal to 400 nm shall not exceed the Class I accessible emission limits for the wavelength range of greater than 1,400 nm but less than or equal to 13,000 nm with a k_1 and k_2 of 1.0 for comparable sampling intervals.
**Class I dual limits: Radiant energy (J) for point sources and integrated radiance (J · cm^{-2} · sr^{-1}) for extended source images.
Source: Bureau of Radiological Health, Reference 1.

Table 7-1B. Class II Accessible Emission Limits for Laser Radiation.

Wavelength (nanometers)	Emission Duration (seconds)	Class II—Accessible Emission Limits
>400 but ≤700	>2.5 × 10⁻¹	$1.0 \times 10^{-3} k_1 k_2 t$ J

Source: Bureau of Radiological Health, Reference 1.

for Class II, III and IV lasers. Table 7-2B gives selected numerical solutions for k_1 and k_2.

Class I lasers are incapable of emitting hazardous laser radiation for any viewing or normal operating conditions. Thus, Class I lasers are exempt from any controls.

Table 7-1C. Class III Accessible Emission Limits for Laser Radiation.

Wavelength (nanometers)	Emission Duration (seconds)	Class III—Accessible Emission Limits
>250 but ≤400	≤2.5 × 10⁻¹	$3.8 \times 10^{-4} k_1 k_2$ J
	>2.5 × 10⁻¹	$1.5 \times 10^{-3} k_1 k_2 t$ J
>400 but ≤1,400	>1.0 × 10⁻⁹ to 2.5 × 10⁻¹	$10 k_1 k_2 t^{1/3}$ J cm⁻² to a maximum value of 10 J cm⁻²
	>2.5 × 10⁻¹	$5.0 \times 10^{-1} t$ J
>1,400 but ≤13,000	>1.0 × 10⁻⁹ to 1.0 × 10¹	10 J cm⁻²
	>1.0 × 10¹	$5.0 \times 10^{-1} t$ J

Source: Bureau of Radiological Health, Reference 1.

Table 7-2A. Values of Wavelength-Dependent Correction Factors k_1 and k_2.

Wavelength (nanometers)	k_1	k_2		
250 to 302.4	1.0	1.0		
>302.4 to 315	$10^{[(\lambda - 302.4)/5]}$	1.0		
>315 to 400	330.0	1.0		
>400 to 700	1.0	1.0		
>700 to 800	$10^{[(\lambda - 700)/515]}$	if: $t \leq \dfrac{10100}{\lambda - 699}$ then: $k_2 = 1.0$	if: $\dfrac{10100}{\lambda - 699} < t \leq 10^4$ then: $k_2 = \dfrac{t(\lambda - 699)}{10100}$	if: $t > 10^4$ then: $k_2 = \dfrac{\lambda - 699}{1.01}$
>800 to 1,060	$10^{[(\lambda - 700)/515]}$	if: $t \leq 100$	if: $100 < t \leq 10^4$	if: $t > 10^4$
>1,060 to 1,400	5.0	then: $k_2 = 1.0$	then: $k_2 = \dfrac{t}{100}$	then: $k_2 = 100$
>1,400 to 1,535	1.0	1.0		
>1,535 to 1,545	$t \leq 10^{-7}$ $k_1 = 100.0$ $t > 10^{-7}$ $k_1 = 1.0$	1.0		
>1,545 to 13,000	1.0	1.0		

Note: The variables in the expressions are the magnitudes of the sampling interval (t), in units of seconds, and the wavelength (λ), in units of nanometers.

Source: Bureau of Radiological Health, Reference 1.

Table 7-2B. Selected Numerical Solutions for k_1 and k_2.

Wavelength (nanometers)	k_1	k_2				
		$t \leq 100$	$t = 300$	$t = 1000$	$t = 3000$	$t \geq 10,000$
250	1.0					
300	1.0					
302	1.0					
303	1.32					
304	2.09					
305	3.31					
306	5.25					
307	8.32					
308	13.2					
309	20.9			1.0		
310	33.1					
311	52.5					
312	83.2					
313	132.0					
314	209.0					
315	330.0					
400	330.0					
401	1.0					
500	1.0					
600	1.0					
700	1.0					
710	1.05	1	1	1.1	3.3	11.0
720	1.09	1	1	2.1	6.3	21.0
730	1.14	1	1	3.1	9.3	31.0
740	1.20	1	1.2	4.1	12.0	41.0
750	1.25	1	1.5	5.0	15.0	50.0
760	1.31	1	1.8	6.0	18.0	60.0
770	1.37	1	2.1	7.0	21.0	70.0
780	1.43	1	2.4	8.0	24.0	80.0
790	1.50	1	2.7	9.0	27.0	90.0
800	1.56	1	3.0	10.0	30.0	100.0
850	1.95	1	3.0	10.0	30.0	100.0
900	2.44	1	3.0	10.0	30.0	100.0
950	3.05	1	3.0	10.0	30.0	100.0
1000	3.82	1	3.0	10.0	30.0	100.0
1050	4.78	1	3.0	10.0	30.0	100.0
1060	5.00	1	3.0	10.0	30.0	100.0
1100	5.00	1	3.0	10.0	30.0	100.0
1400	5.00	1	3.0	10.0	30.0	100.0
1500	1.0					
1540	100.0*			1.0		
1600	1.0					
13000	1.0					

*The factor $k_1 = 100.0$ when $t \leq 10^{-7}$, and $k_1 = 1.0$ when $t > 10^{-7}$.

Note: The variable (t) is the magnitude of the sampling interval in units of seconds.

Source: Bureau of Radiological Health, Reference 1.

Note that Table 7-1A also provides accessible emission limits for collateral (i.e., nonlaser) radiation emitted from laser products.

Class II Laser

Two groupings are defined for Class II lasers:

- Visible lasers that could provide human access to laser radiation in excess of the accessible emission limits of Class I, but not in excess of the accessible emission limits of Class II of Table 7-1B, in the wavelength range of >400 nm but ≤ 700 nm for emission durations >0.25 second.
- Lasers that do not provide human access to laser radiation in excess of the accessible emission limits of Class I for any other combination of emission duration and wavelength range.

Visible Class II lasers have sufficient output power to produce retinal injury when stared at for a long time, and therefore require certain precautions to be exercised; however, they do not have sufficient output power to cause injury accidentally.

Class III Laser

This class laser could provide human access to laser radiation in excess of the accessible emission limits of Class I and, if applicable, Class II, but does not provide human access to laser radiation in excess of the accessible emission limits of Class III of Table 7-1C for any combination of emission duration and wavelength range. Class III lasers are separately designated as Class IIIa or Class IIIb (for labeling requirements), based on the following criteria:

- Class IIIa—so classified because of the emission of accessible visible laser radiation for emission durations greater than 3.8×10^{-4} second and in the wavelength range of >400 nm but ≤ 700 nm, with an irradiance ≤ 2.5×10^{-3} W · cm^{-2} and with a radiant power ≤ 5.0×10^{-3} W.
- Class IIIb—each Class III laser other than that group described above.

Class IIIa (visible) lasers are incapable of damaging the eye because of the person's normal aversion response to bright light, unless binoculars or similar optical instruments are used, or unless the radiation is stared at for a long time. However, Class IIIb lasers are capable of causing accidental injuries by exposure from the direct or specularly reflected beam. Diffuse laser beam reflections are not hazardous, but may be so if focused to the eye with optical instruments. Many military laser rangefinders and designators fall into the Class IIIb category.

Class IV Laser

This class of laser could provide human access to laser radiation in excess of the accessible emission limits of Class III lasers. (See Table 7-1C.)

Class IV lasers can produce a hazardous direct or specularly reflected beam. Also, a potential fire and skin burn hazard exists, as does the possibility of hazardous diffuse reflections occurring.

Class V Laser (Enclosed Laser)

The American National Standard Institute[3] has eliminated the Class V category in order to make its classification system compatible with that of the BRH.

CENTRAL BEAM IRRADIANCE OR RADIANT EXPOSURE[4]

The laser specification often fails to provide a measure of beam irradiance or radiant exposure, which are parameters needed for laser classification and MPE assessments. For single mode lasers having a Gaussian beam profile, the central beam values may be obtained from the beam diameter specified at $1/e$ points and the beam radiant power or energy. The equations are:

$$E_0 = \frac{4\Phi}{\pi a^2} = \frac{1.27\Phi}{a^2} \qquad \text{(Eq. 7-1)}$$

$$H_0 = \frac{4Q}{\pi a^2} = \frac{1.27Q}{a^2} \qquad \text{(Eq. 7-2)}$$

where

E_0 = emergent beam irradiance at zero range, measured in $W \cdot cm^{-2}$ for CW lasers

H_0 = emergent beam radiant exposure at zero range, measured in $J \cdot cm^{-2}$ for pulsed lasers

Φ = total radiant power output of a CW laser, or average radiant power of a repetitively pulsed laser, measured in watts

Q = total radiant energy output of a pulsed laser, measured in joules

a = diameter of emergent laser beam, measured in cm

In many cases the beam divergence or diameter is specified at $1/e^2$ points rather than $1/e$ points. Equations 7-1 and 7-2 would then provide average values rather than central beam values. It then becomes necessary to multiply these specifications by $\sqrt{2}$ to obtain the $1/e$ values.

EXAMPLES OF LASER CLASSIFICATION

Examples 7-1 to 7-4, which follow, are classification problems that would be of particular interest to the laser user. Problem C-1 of Appendix C follows a suggested BRH format shown in Table 10-1 of Chapter 10 (Public Laws) and approaches laser classification from a manufacturer's viewpoint or requirement.

Example 7-1. Classify a Q-switched ruby laser of wavelength 694.3 nm that has an output peak power specified by the manufacturer as 10 MW, a pulse duration of 25 nsec, and a $\frac{7}{8}$-inch laser rod diameter.

1) $\lambda = 694.3$ nm
$p = 10$ MW $= 10 \times 10^6$ watts
$t = 25$ nsec $= 25 \times 10^{-9}$ second
Laser rod diameter $= \frac{7}{8}$ inch $= (\frac{7}{8}$ inch$)(2.54$ cm/inch$) = 2.22$ cm

2) $Q = p \cdot t = (10 \times 10^6 \, W)(25 \times 10^{-9} \, \text{second})$

$= 0.25 \, W \cdot \text{second} = 0.25 \, J$

3) The emergent beam diameter, a, is not specified. Thus, H_0 can only be estimated. The value of a is always smaller than the laser rod diameter.

$$\therefore \quad a < 2.22 \text{ cm}$$

4) Hence, the emergent beam radiant exposure (H_0) can be no less than (from Eq. 7-2)

$$H_0 = \frac{1.27Q}{a^2} = \frac{(1.27)(0.25 \text{ J})}{(2.22 \text{ cm})^2}$$

$$= 0.0644 \text{ J} \cdot \text{cm}^{-2}$$

5) From Table 7-1C for $\lambda = 694.3$ nm (i.e., >400 but $\leq 1,400$ nm) and for emission duration of 25×10^{-9} second (between $>1.0 \times 10^{-9}$ and 2.5×10^{-1} second), the Class III Accessible Emission Limit is given by:

$$H = 10k_1 k_2 t^{1/3} \text{ J} \cdot \text{cm}^{-2}$$

From Table 7-2A for $\lambda = 694.3$ nm (i.e., between >400 and 700 nm), or from Table 7-2B,

$$k_1 = 1 \quad \text{and} \quad k_2 = 1$$

Thus, the Class III Accessible Emission Limit is:

$$H = 10(1)(1)(25 \times 10^{-9})^{1/3} \text{ J} \cdot \text{cm}^{-2}$$

$$= 29.2 \times 10^{-3} \text{ J} \cdot \text{cm}^{-2}$$

$$= 0.0292 \text{ J} \cdot \text{cm}^{-2}$$

From step (4) above, the emergent beam radiant exposure (H_0) is 0.0644 J $\cdot$ cm^{-2} which exceeds the Class III limit by more than two times. Thus, we have a Class IV laser.

Example 7-2. A tunable laser is capable of emitting between 500 and 1,500 nm at an output of 25 mJ, has a 5-mm-diameter beam and an emission duration of 1 μsec. A manual switching arrangement has been devised to fire separately at wavelengths of 500, 1,000 and 1,300 nm. Classify the laser.

1) $Q = 25$ mJ $= 25 \times 10^{-3}$ J
$a = 5$ mm $= 0.5$ cm
$t = 1$ μsec $= 1 \times 10^{-6}$ second
$\lambda_1 = 500$ nm; $\lambda_2 = 1,000$ nm; $\lambda_3 = 1,300$ nm

2) From Eq. 7-2, the emergent beam radiant exposure is

$$H_0 = \frac{1.27Q}{a^2} = \frac{(1.27)(25 \times 10^{-3})}{(0.5)^2}$$

$$= 0.127 \text{ J} \cdot \text{cm}^{-2}$$

3) From Table 7-2A:

for $\lambda_1 = 500$ nm: $k_1 = 1.0$ and $k_2 = 1.0$

for $\lambda_2 = 1,000$ nm: $k_1 = 10^{[(\lambda - 700)/515]} = 10^{[(1000 - 700)/515]}$

$$= 10^{.583} = 3.82$$

$k_2 = 1.0$ for $t \leq 100$ (e.g., $t = 1 \times 10^{-6}$ sec)

for $\lambda_3 = 1,300$ nm: $k_1 = 5.0$

$k_2 = 1.0$ for $t \leq 100$ (e.g., $t = 1 \times 10^{-6}$ sec)

4) From Table 7-1C for $t > 1.0 \times 10^{-9}$ to 2.5×10^{-1} (e.g., $t = 1 \times 10^{-6}$ sec) Class III Accessible Emission Limit is:

$10k_1 k_2 t^{1/3}$ J $\cdot$ cm^{-2} to a maximum value of 10 J $\cdot$ cm^{-2}

for $\lambda_1 = 500$ nm: $10 (1.0)(1.0)(1 \times 10^{-6})^{1/3}$ J $\cdot$ cm^{-2}

$$= 0.1 \text{ J} \cdot \text{cm}^{-2}$$

for $\lambda_2 = 1,000$ nm: $10 (3.82)(1.0)(1 \times 10^{-6})^{1/3}$ J $\cdot$ cm^{-2}

$$= 0.382 \text{ J} \cdot \text{cm}^{-2}$$

for $\lambda_3 = 1,300$ nm: $10 (5.0)(1.0)(1 \times 10^{-6})^{1/3}$ J $\cdot$ cm^{-2}

$$= 0.50 \text{ J} \cdot \text{cm}^{-2}$$

5) Although H_0 does not exceed the Class III accessible emission limits at 1,000 or 1,300 nm, Class III limits are exceeded at 500 nm (i.e., 0.127 J $\cdot$ cm^{-2} vs. limit of 0.1 J $\cdot$ cm^{-2}). Thus, we have a Class IV laser.

Example 7-3.*   A one-watt argon laser operating at 514.5 nm is intended to be used in a communications link. Under what condition would the emergent beam not be considered a skin hazard? Also, determine the hazard classifications that would apply.

1) Φ = 1 watt
   λ = 514.5 nm

2) Since the radiant power output (Φ) is >0.5 watt, a potential skin burn hazard exists.

3) From Table 6-4 of Chapter 6, for λ = 514.5 nm (i.e., between 400 and 1,400 nm) the skin exposure MPE = 0.2 W $\cdot$ cm^{-2}, for exposure durations between 10 and 3 × 10^4 seconds.

4) The beam would have to be made sufficiently large or broadened to reduce the emergent beam irradiance (E_0) to less than 0.2 W $\cdot$ cm^{-2}. From the equation

$$E_0 = \frac{4\Phi}{\pi a^2},$$

$$0.2 \text{ W} \cdot \text{cm}^{-2} = \frac{(4)(0.5 \text{ W})}{\pi a^2}$$

$$a = \sqrt{\frac{(4)(0.5)}{(3.14)(0.2)}} = 1.78 \text{ cm}$$

Therefore, the beam diameter must be greater than 1.78 cm to avoid a potential skin hazard problem.

5) The laser could fall into one of three classifications.

   • Class I, if the entire laser beam path were enclosed.
   • Class IV, if >0.5 watt emerged from the laser system as an unenclosed beam.
   • Class III, if after passing through beam-forming optics the total optical power in the beam were <0.5 watt.

Example 7-4.*   Classify a 632.8-nm visible laser (He-Ne) utilized as a remote control switch. The laser is electronically pulsed with a

*Adapted from Reference 4.

1-mW peak power output, a pulse duration of 0.1 second and a beam diameter of 1 cm. The recycle time of the laser is 5 seconds.

1) $\lambda = 632.8$ nm
$p = 1$ mW $= 1 \times 10^{-3}$ watt
$t = 0.1$ second
$a = 1$ cm

2) $$Q = p \cdot t \ (10^{-3}\,\text{W}) \cdot (0.1 \text{ second})$$
$$= 10^{-4}\,\text{W} \cdot \text{second}$$
$$= 10^{-4}\,\text{J for each pulse}$$

3) Since the recycle time of the laser is 5 seconds, the PRF = 0.2 Hz.

4) Since the device is pulsed with an exposure duration of 0.1 second, then, from Figure 6-5 of Chapter 6, the applicable MPE for intrabeam viewing is $3.2 \times 10^{-4}\,\text{J} \cdot \text{cm}^{-2}$.

5) The emergent beam radiant exposure per pulse is

$$H_0 = \frac{1.27Q}{a^2} = \frac{(1.27)(10^{-4})}{(1)^2}$$
$$= 1.27 \times 10^{-4}\,\text{J} \cdot \text{cm}^{-2}$$

This is less than half the MPE, which indicates the laser is Class I, exempt, if repeated exposure is not considered possible.

6) Since biologic data and MPEs for exposures repeated at PRFs less than 1 Hz are lacking, the exposures should be considered linearly additive. Thus, at least two exposures ($2 \times 1.27 \times 10^{-4}\,\text{J} \cdot \text{cm}^{-2} < 3.2 \times 10^{-4}\,\text{J} \cdot \text{cm}^{-2}$) are allowable.

THE LASER ENVIRONMENT

As noted at the beginning of the chapter, a second key element in evaluation of the laser beam hazard is the laser environment. Consideration of the environment, of course, is necessary only when the laser classification indicates a potential for injury. Given that a potentially hazardous laser is being used, the probability of exposure is relatable to its usage indoors or outdoors. For example, lasing in a light-tight laboratory enclosure restricts the potential hazard to those individuals in the enclosure. On the other hand, laser usage against a target from an aircraft at a military test range poses a potential hazard over a much expanded airspace.

Some considerations common to both indoor and outdoor environments must be addressed in evaluating the laser beam hazard, so that the appropriate control measures can be taken (see Chapter 8, Control of Laser Radiation Hazard). These follow:

- Thought must be given to the possibility of individuals entering an environment where lasing is taking place. Concern could be minimal for a highly restricted laboratory. A large factory building or aircraft hangar with many doors might be of increasing concern, while airborne lasing covering a vast airspace would be of great concern.
- An assessment of the stability of the laser beam path or equipment platform must be made. Perhaps constraints are possible and necessary to limit lateral and elevation movement of the platform.
- The potentially hazardous laser beam path(s) or zone(s) must be identified. Clearly, the problem is much simpler for a non-movable laser in a confined area, while much more difficult with an aircraft-mounted and slewable laser.
- The potential hazard caused by specular reflections requires review and identification where possible. Shiny, flat metallic surfaces are usually of concern, but are more easily controllable in confined areas. Larger or outdoor areas for lasing usually present a greater problem because of the increased probability of more specular surfaces, such as unseen objects, still-water ponds, ice, and so on. Also, it may be difficult to remove or render diffuse some specular objects.
- Ascertain whether diffuse reflections are possible or exist during lasing. This can be of concern for Class IV lasers, but is no problem for Class III lasers unless the laser beam is focused to the eye by an optical instrument such as binoculars.
- Examination of the intended laser usage in the laser environment may also be required for a satisfactory beam hazard evaluation. For example, military tactical exercises whereby rifle laser designators are used to illuminate targets for laser-

guided weapons are potentially hazardous, as are airborne firings against ground targets. Indoor usage in a laboratory requiring frequent configuration changes and firings is also a hazard requiring evaluation.

For outdoor laser usage over extended ranges it is necessary to calculate or measure the hazardous ranges for the direct or specularly reflected laser beam for all paths or zones that can be hazardous. Radiant exposure or beam irradiance as a function of range is calculated using Eqs. 6-10 and 6-11, which follow:

$$E = \frac{\Phi e^{-\mu r}}{\pi \left[\dfrac{a + r\phi}{2} \right]^2} = \frac{1.27 \Phi e^{-\mu r}}{(a + r\phi)^2}$$

$$H = \frac{Q e^{-\mu r}}{\pi \left[\dfrac{a + r\phi}{2} \right]^2} = \frac{1.27 Q e^{-\mu r}}{(a + r\phi)^2}$$

THE PERSONNEL PRESENT IN THE LASER ENVIRONMENT

The third element required to evaluate the laser beam hazard is consideration of the personnel present in the laser environment, and includes user personnel as well as any other individuals in the proximity of the lasing. Concern with personnel in the laser environment is naturally required only for laser classes that provide a potential for injury. Some important personnel considerations that bear addressing in order to evaluate this aspect of the laser beam hazard are:

- Skill of the individual who is in charge of assuring a hazard-free laser operation (i.e., Laser Safety Officer or other individual).
- The presence of individuals who may be unable to read or comprehend warning signals, such as children.
- Adequacy of control of the location and number of individuals, to minimize inadvertent exposures from the direct laser beam or reflections.
- Knowledge of individuals in the laser environment that potentially injurious laser radiation may exist.
- The quality of laser safety indoctrination and training of the

individuals directly associated with the laser exercise, as well as their experience.

- The maturity and good judgment of the laser user (e.g., can the person(s) be trusted to comply with established hazard control procedures?).
- Assurance that no individuals with physical impairments such as defective hearing or eyesight serve any critical safety function (e.g., posted safety guard for limiting vehicular/personnel access into a hazardous firing area).

REFERENCES

1. "Laser Products Performance Standards," Federal Register, Vol. 40 No. 148, pp. 32252–32266, U.S. Dept. of Health, Education and Welfare, Food and Drug Administration, Bureau of Radiological Health (BRH), Washington, D.C., 31 July 1975.
2. "Guide For Submission of Information on Lasers and Products Containing Lasers Pursuant to CFR 1002.10 and 1002.12," U.S. Dept. of Health, Education, and Welfare, Public Health Service, Food and Drug Administration, Bureau of Radiological Health, Division of Compliance, Rockville, Maryland, July 1976.
3. "American National Standard for the Safe Use of Lasers, Z 136.1-1976," American National Standards Institute, New York, 1976.
4. "Control of Hazards to Health from Laser Radiation," U.S. Dept. of the Army Technical Bulletin TB-MED-279, Washington, D.C., May 1975.

BIBLIOGRAPHY

1. "NATC Instruction 5100.2, Laser Safety; responsibilities and minimum safety procedures for," Department of the Navy, Naval Air Test Center, Patuxent River, Maryland, 5 Mar. 1973.

8

Control of Laser Radiation Hazard

INTRODUCTION

The major objective in devising and implementing control measures is to minimize the probability of exposing the eye and the skin to potentially dangerous laser radiation.

Hazard control measures are influenced by the classification of the laser being used. Indeed, control measures that follow will, in part, be addressed on the basis of laser classification. However, the laser classification scheme relates specifically to the laser device itself and its potential hazard based on operating characteristics. Therefore, a more comprehensive approach necessitates consideration of the environment and the manner in which the laser is used, as well as personnel factors.

Many of the recommended precautions and control measures are directly related to the following:

- Engineering controls
- Administrative procedures
- Control of the environment in which the laser is operating

For an indoor environment, engineering controls tend to play the predominant role. On the other hand, for an outdoor environment, administrative controls often provide the only reasonable approach to minimizing the laser beam hazard.

Probably the best safety rule is to appreciate that a potential hazard exists with many lasers, and to combine this appreciation with good common sense. It is likewise clear, that an informed laser user will in general be a safer user. Thus, provision for some minimal laser safety training program makes good sense.

Laser users should select systems that provide the lowest power or energy levels necessary to fullfill their requirements.

It is the responsibility of a designated Laser Safety Officer (LSO) to develop and implement effective safety rules and procedures for controlling the laser hazards. (See Chaper 11, Laser Safety Program, for complete LSO functions and responsibilities.) A Laser Range Safety Officer or LRSO is the individual designated by the U.S. Army for implementing safe operating procedures at laser ranges.[1] At the Grumman Aerospace Corporation, the equivalent individual for the A-6E TRAM aircraft ground and flight development tests is called a Laser Safety Coordinator or LSC.

The voluntary ANSI[2] safety standard indicates a general lessening of laser control measures when usage is for diagnostic, therapeutic or medical research purposes, and at the direction of qualified professionals engaged in the healing arts. However, protective control measures apply to those administering the laser radiation.

The laser equipment design, labeling and safety instructions all play a role in controlling the laser radiation hazard. These aspects and others, governing the manufacture of laser products per the Bureau of Radiological Health's Laser Products Performance Standards,[3] are covered in Chapter 10, Public Laws.

This chapter is divided into the following sections:

- Class I—Exempt Laser Control Measures
- Class II—Low Power Laser Control Measures
- Class III—Medium Power Laser Control Measures
- Class IV—High Power Laser Control Measures
- Class V—Enclosed Laser Control Measures (extinct category)
- Infrared Lasers—Special Control Measures
- Ultraviolet Lasers—Special Control Measures
- Field and Airborne Lasers—Control Measures
- Laser Protective Eyewear

- Warning Signs and Labels
- Alteration of Output Power or Operating Characteristics of Laser

CLASS I—EXEMPT LASER CONTROL MEASURES

Since Class I lasers are incapable of producing damaging radiation levels, no control measures are required, including the use of warning labels. Good practice and good common sense dictate that unnecessary and prolonged gazing at the laser beam must be avoided.

CLASS II—LOW POWER LASER CONTROL MEASURES

Class II visible lasers are incapable of causing eye injury accidentally, but can produce retinal injury with continuous intrabeam viewing. Thus, the single most important control measure is to avoid staring at the beam for a long time.

Secondly, suitable warning labels should be placed prominently on the laser housing and/or control panel. The Bureau of Radiological Health (BRH)[3] requires the manufacturer to affix a "caution" label, as shown in Figure 8-1.

Figure 8-1. BRH label for Class II helium-neon laser.

CLASS III—MEDIUM POWER LASER CONTROL MEASURES

Class III lasers are potentially hazardous to the eye, and require control measures to minimize the possibility of injury to that organ. Two categories of Class III lasers are specified by the BRH,[3] namely, Class IIIa and Class IIIb.

Class IIIa lasers emit visible radiation that is incapable of injuring the eye within the duration of the blink or aversion response (~0.25 second). Injury is, however, possible using binoculars or similar optical devices, or by continued staring at the beam. The BRH characterizes Class IIIa lasers as having emission durations $>3.8 \times 10^{-4}$ second, wavelength >400 nm but ≤ 700 nm, irradiance $\leq 2.5 \times 10^{-3}$ $W \cdot cm^{-2}$ and a radiant power $\leq 5.0 \times 10^{-3}$ watts.

Accidental injury to the eye is possible with Class IIIb lasers, if the beam is viewed directly or from a specularly reflected beam.

Ten categories of control measures for medium power lasers will be covered. It generally follows that these same control measures would also be applicable for high-powered Class IV lasers, with additional safeguards needed for the latter type classification because of its higher power. The ten categories follow:

- Education and training of laser users
- Engineering controls
- Administrative controls
- Laser controlled area
- Laser operators
- Spectators
- Alignment procedures
- Optical viewing devices
- Eye protection
- Equipment labeling

While many of the control measures for Class III and IV lasers apply for outdoor usage in a field or an airborne environment, there are special aspects and approaches necessitating additional treatment, which will be handled separately.

Education and Training of Laser Users

Laser users must be properly educated as to the potential hazards involved while operating these devices. Thus, the laser safety training of user personnel is an essential element in the implementation of successful laser radiation hazard control measures. (See Chapter 11, Laser Safety Program, for approaches to laser safety training.)

Further, only experienced persons should be permitted to operate lasers.

Engineering Controls

Implementation of suitably engineered safety mechanisms and techniques should be made an integral part of the laser operation. Examples of such engineering controls follow:

- Beam stops or backstops
- Reduction of laser beam intensity
- Beam enclosure
- Beam shutters
- Firm laser mount
- Controlled beam path
- Strong illumination of enclosed laser area
- Detailed operating and safety procedures
- Removal or elimination of specular reflectors
- Diffuse characteristic of target
- Beam path not at eye level

Where feasible, the beam should be halted at the end of its useful path. This might be achieved by constructing an oversized beamstop surrounding the target. For an outdoor operation, substantial backstops might take the form of densely wooded areas, earth mounds or hills.

Situations may arise where a reduction of the laser beam intensity to less hazardous levels is allowable and even desirable, with no loss of purpose resulting. At least two approaches are feasible. The first would be to insert an absorbing filter in the path of the beam. Alternately, a beam-enlarging method for increasing the beam diameter using a pair of lenses is possible. The first lens spreads the beam, while the second lens suitably placed would serve to recollimate the widened beam. The resultant decrease in power or energy density reduces the hazard posed by beam viewing.

Enclosing the laser beam path as often as possible serves to reduce the hazard considerably. A simple and inexpensive cardboard box can be used as a beam enclosure for classroom experiments or demonstrations. (See Chapter 12, Safety in Classroom Laser Use).

Many lasers are built for sustained operation, with useful life shortened by frequent enabling and disabling of the device. An easy way of overcoming this problem is to use a shutter, which can be a simple mechanical device fitted over the aperture of the laser. Keeping the beam enclosed while the shutter covers the aperture reduces the laser beam hazard. The shutter should completely stop the laser beam. A black nonreflective material is desirable.

Mounting the laser on a firm support such as a tripod structure will help to assure that the beam travels along its intended path.

It is possible to ensure a controlled beam path with many laser systems. One relatively simple approach would be to provide physical stops for constraining elevation and azimuth angle movement. For more elaborate systems where mount movement is computer-controlled, software can be readily designed and incorporated to provide automatic laser firing disabling when prescribed azimuth or elevation limits are exceeded.

For an enclosed laser area it is desirable to illuminate the area as brightly as possible. This illumination constricts the observer's pupil size and limits the amount of laser radiation that would fall upon the retina if an intrabeam viewing situation arose.

Suitably detailed operating procedures with full safety factors considered should be prepared before operation of the laser. They are particularly important prior to first-time operation and also serve as a necessary operating guide when more than one person or group is involved in the laser operation. It becomes almost a mandatory requirement for more complex laser systems where procedures naturally tend to become more complex. Although the preparation and usage of laser operating and safety procedures may be considered an administrative control, and this is certainly reasonable, the preparation and initial implementation of such procedures normally falls into the engineering domain. For this reason, this area has been included as an engineering rather than an administrative control. A word of caution is in order. Laser safety procedures can sometimes become an implementation nightmare if good engineering judgment and common sense are not exercised.

Removal of specular or mirror like surfaces from the vicinity of the laser beam path should be done when possible. Likewise, firings should be restricted against known specular objects. Where circumstances make it difficult or impossible to remove specular surfaces from the firing area, then painting such surfaces to impart a diffuse or beam-spreading property is quite important. Use of an inexpensive flat dull black paint is very effective.

The laser beam will travel outward from the laser until it is absorbed or reflected. The intended target should be a diffuse, absorbing material to prevent potentially dangerous specular reflections. Black foam rubber material or black ink on blotter paper makes a good diffuse target. Sandblasted aluminum sheets have been found to work quite effectively as a diffusing target material for the near infrared emissions on the A-6E TRAM project.

When possible, it should be attempted to set up the laser beam path so that it is outside normal standing or sitting eye level position, i.e., below three feet or above six or so feet. Classroom laser experiments or demonstrations would be concerned with this consideration.

Administrative Controls

Administrative control measures can help to minimize the laser beam hazard. The almost unlimited possibilities for laser usage make it impossible to foresee and document every useful administrative control measure. However, the listing that follows provides a good baseline to build upon.

- The laser beam should not be aimed at people, particularly their eyes.
- Avoid looking into the primary laser beam.
- Avoid looking at visible laser beam reflections, as these too have the potential for causing retinal burns.
- Avoid looking at an exposed laser pump source.
- Safety eyewear designed to filter out the specific laser frequencies should be worn whenever risk of exposure to hazardous levels exist.
- The laser should not be left unattended if the possibility exists that unauthorized and unqualified personnel may use it.

- Good housekeeping practices should be used to ensure that no specular objects are discarded in the laser beam path.
- Remove watches, rings, bracelets and other shiny jewelry, as hazardous reflections are possible.
- When possible, all personnel should be cleared from the anticipated beam path.
- For indoor installations, doors should be locked during laser usage to keep out unwanted personnel.
- To the extent feasible, lasers should be operated in well-controlled areas, for example, in a closed room with covered or filtered windows and controlled access.
- Prior to operating the laser all personnel, particularly visitors, should be warned of the potential hazard. A reminder that they have but one set of eyes is not out of place.
- Conspicuous signs indicating that the laser is in operation and that it could be dangerous should be prominently placed both inside and outside the work area, and on doors providing access to the area.
- Particular caution should be exhibited in working with lasers that operate at invisible spectral regions.
- Directing the laser beam at, or tracking, nontarget vehicles or aircraft is strictly forbidden.
- Any afterimages may be indicative of eye damage and should be reported to a doctor, preferably an ophthalmologist experienced with retinal burns.

Laser Controlled Area

For Class III laser usage, strong consideration should be given to operating the laser device in a controlled area. For high power Class IV lasers it is mandatory. A full discussion relative to the laser controlled area as a control measure will be dealt with under "Class IV—High Power Laser Control Measures."

Laser Operators

Only authorized operators should fire laser systems. However, rules and regulations governing certification of laser operators are not federally controlled, while state regulations vary or are non-existent. New York Industrial Code Rule 50[4] requires a Certificate of Competence for mobile laser operators (i.e., a laser used or operated outside a laser installation). This rule is however waived for research and development programs and for professional engineers or land surveyors licensed to practice in New York State.

Spectators

In general, spectators should be severely restricted or prohibited from entering laser controlled areas. Where spectators are permitted, it should be by authorization of suitable supervisory personnel, with the necessary protective measures taken. In addition, it should be made quite clear that a potential beam hazard exists.

Alignment Procedures

During the alignment of laser optical components, which include mirrors, beam deflectors, and expanders, lenses, etc., the techniques used should ensure that the eyes are not exposed to the primary beam or hazardous reflections of the primary beam.

Optical Viewing Devices

Optical viewing devices such as binoculars, microscopes and telescopes can increase the eye hazard when one is viewing a laser beam. If optical devices are used for viewing, protective features such as filters and suitable interlocks should be utilized to ensure that eye exposure levels are below MPE values for irradiation of the eye.

Eye Protection

If it has been determined that all engineering and administrative control measures fail to eliminate the possibility of exposures exceeding applicable MPE levels for the eye, then suitable eye protection devices must be worn.

Equipment Labeling

Warning labels should be placed on a conspicuous spot on the laser housing and/or control panel. The Bureau of Radiological Health (BRH)[3] specifies labeling requirements based on two categories of Class III lasers, namely, Class IIIa and Class IIIb. See Figure 8-2 for the Class IIIa BRH required warning label. Position 1 should read "LASER RADIATION—DO NOT STARE INTO BEAM OR VIEW DIRECTLY WITH OPTICAL INSTRUMENTS." Position 3 should read "Class IIIa LASER PRODUCT." Position 2 is reserved for radiation output information (same for Class II, IIIb and IV lasers). Using appropriate units, maximum output of laser radiation, pulse duration when appropriate and laser medium or emitted wavelength are required.

The required BRH warning label for Class IIIb lasers is shown in Figure 8-3. Position 1 should read "LASER RADIATION—AVOID

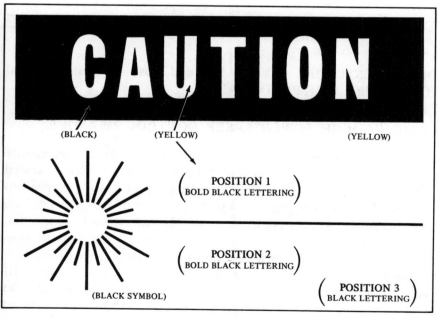

Figure 8-2. BRH Class IIIa warning label.

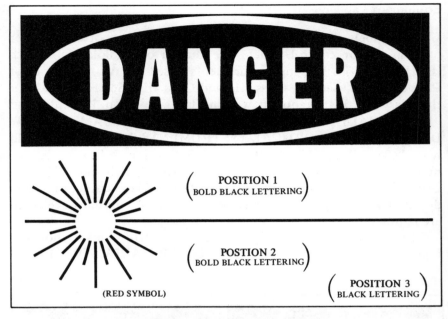

Figure 8-3. BRH Class IIIb warning label.

DIRECT EXPOSURE TO BEAM," and Position 3 should read "CLASS IIIb LASER PRODUCT."

CLASS IV—HIGH POWER LASER CONTROL MEASURES

High-powered lasers are the most hazardous lasers in that the chance of injury and its seriousness increase commensurate with increased power or energy output. The possibility of generating dangerous specular reflections is greater, as is the prospect of hazardous diffuse reflections. Aside from the eye and skin hazard, Class IV lasers present a potential fire hazard. Because of the aforementioned considerations, more stringent hazard control measures are required.

The previous section on control measures to be used for Class III lasers cited and discussed ten such categories, which are by and large applicable to Class IV high power lasers. The laser controlled area and engineering controls will be discussed, with additional requirements for engineering controls over that already cited for Class III lasers.

See Figure 8-3 for the BRH required warning label. Position 1 should read "LASER RADIATION–AVOID EYE OR SKIN EX-POSURE TO DIRECT OR SCATTERED RADIATION." Position 3 should read "CLASS IV LASER PRODUCT."

Laser Controlled Area

High-powered lasing should be restricted to a controlled area, suitably administered and designed in order to protect individuals from exposure to the laser radiation. Such an area is referred to as a laser controlled area, and may be located either indoors or outdoors, depending on the laser application. While laser usage objectives may vary, the responsibility for laser safety in a controlled area must rest with a single individual, namely, the Laser Safety Officer or his duly appointed representative, with the overall operation governed by written safety procedures. Approval and surveillance of these procedures rests with the Laser Safety Officer.

For indoor operations, safety latches, interlocks or persons used as door guards should be utilized to ensure that unexpected and un-protected individuals do not enter a laser controlled area, and that radiation is not emitted when doors are opened. The facility should be light-tight, with windows or other optical openings covered or restricted in such a way as to prevent emission of hazardous laser radiation levels. The capability for rapid departure and entry to the area under emergency conditions is necessary. A "panic switch" should be made available for deactivating the laser under emergency conditions.

For outdoor laser installations, the laser beam should not be directed against nontarget vehicles or aircraft, nor should it be fired out of a controlled air or ground space within known hazardous distances. Radio communications, guards and control of ground and air traffic are some of the administrative controls frequently used in outdoor laser controlled areas. Physical barriers at roadway accesses are often useful.

There are some items applicable to both indoor and outdoor laser controlled areas. Access to such areas should be allowed only via appropriate authorization. All personnel should wear adequate eye protection, and skin protection if a skin hazard exists, whenever the laser is capable of emission. If a potentially serious skin or fire hazard exists, then an appropriate safety shield must separate the laser beam from the personnel. Appropriate warning signs should be placed at key locations, such as doors, roads, and so on. Where laser power or energy levels are sufficiently high, the use of diffusing, absorbing and fire-resistant backstops and targets may be required. The beam path(s) should be cleared of all specular surfaces capable of producing potentially hazardous reflections. If removal is not possible, then a diffuse covering should be used over or on the objects.

For normal maintenance, troubleshooting or training purposes, where protective covers must be taken off and interlock protection circumvented, a temporary laser controlled area should be established. As is the case with the regular laser controlled area, written safety procedures governing the operation during maintenance and servicing must be approved by the Laser Safety Officer. While certain built-in protective features will be overridden, the temporary laser controlled area must still provide for all safety requirements during maintenance, training and servicing.

Engineering Controls

In addition to the engineering controls specified for Class III lasers, which are also applicable to high-powered lasers, the following requirements exist or should be strongly considered:

- Enclosure of beam path
- Remote firing and monitoring
- Warning systems and/or procedures
- Key-switch master interlock
- Protection against accidental firing of pulsed lasers

Enclosure of the whole laser beam path should be implemented for high power lasers whenever possible. This includes enclosing the interaction area, or area in which irradiation of objects by the primary or reflected beams takes place. Interlocks should be used with the enclosures to ensure that the laser cannot operate unless the enclosure is correctly built.

When feasible, remote firing and video monitoring should be used, thus eliminating the need for personnel to be physically present in the same room or in the direct or reflected line of sight of the beam. Real-time video monitoring of the beam footprint is a particularly desirable safety feature for invisible lasers, as well as potentially useful as an engineering analysis tool.

Prior to laser activation, involved personnel should be forewarned of an impending firing. For indoor operations, suitable alarms or warning systems, such as a rotating light that can be seen while wearing protective eyewear, audible sound or a verbal "countdown," are satisfactory approaches. For extended lasing periods, it is often desirable, as well as mandatory, that a conspicuous but not too irritating alarm be maintained, such as a visible rotating light. While the same alarms may also suffice for outdoor operations, simple radio communication might provide the only warning where a broad expanse is involved in the laser exercise, as might be the case at a military lasing range.

For most Class IV lasers, a keyed master interlock or switching device is necessary to ensure that only authorized personnel operate the laser. A removable key should be used and the laser rendered inoperable when the key is removed. This approach is not always feasible for some military applications.

Once charged, optical pumping systems for pulsed lasers can be spontaneously discharged, causing the laser to fire. Thus, the firing circuit design should incorporate protective measures in order to prevent such an occurrence.

CLASS V—ENCLOSED LASER CONTROL MEASURES

The American National Standards Institute has eliminated the Class V category in order to make its classification system compatible with the BRH system. Thus, no control measures exist.

INFRARED LASERS—SPECIAL CONTROL MEASURES

Special attention is required in working with infrared lasers, since the radiation is invisible (as is ultraviolet radiation). The control measures previously enumerated and based on laser classification are still largely applicable, but the following requirements frequently are of particular concern:

- Beam path termination
- Video monitoring
- Filter materials for very high-powered lasers

Beam Path Termination

For Class III lasers, consideration should be given to terminating the beam path with a very absorbent backstop. This approach is usually more readily implemented for indoor operations, although for outdoor operations advantage is frequently taken of natural terrain features for satisfying the beam termination requirement. A dull appearance on some surfaces can be quite misleading relative to reflecting infrared radiation.

For Class IV lasers, the beam path termination ought to be fire-resistant, when this approach is feasible. Inspection of the absorbent should be made on a regular basis, since many materials degrade with usage.

For very high-powered lasers, water-cooled metal reflectors with convex surfaces may be needed to diverge the beam and so reduce the output level to a point where it can be dissipated by the absorbing backstop.[5]

Video Monitoring

Video monitoring of the infrared laser radiation for positive location of the beam path is sometimes a required control measure.

Filter Materials for Very High-Powered Lasers

Filter materials usually utilized as protective beam shields or goggles may be useless with very high-powered lasers if the output levels exceed the material's damage threshold. Further, if output levels are high enough to cause serious skin burns or ignite clothing, techniques should be used to lower the level of radiation intercepted by personnel to a skin safe level and a level not inflammatory for clothing.

ULTRAVIOLET LASERS SPECIAL CONTROL MEASURES

As with infrared radiation, ultraviolet radiation is also invisible, and once again special care must be exercised. The control measures based on laser classification are likewise applicable.

Particular attention must be given to the possibility of producing reactions in the presence of ultraviolet radiation, for example, formation of ozone and skin-sensitizing agents.[2]

FIELD AND AIRBORNE LASERS—CONTROL MEASURES

Numerous laser systems are being used in tactical military ground and airborne applications. Laser ranging, weapon guidance (e.g., "smart" bombs) and target designation are but a few of the increasingly common applications. The laser systems may be vehicle- or aircraft-mounted, tripod-mounted or hand-held. Many of the devices have intrabeam viewing hazard distances of several miles.

In general, the safety guidelines for field laser operation using potentially hazardous lasers are built around:

- Limiting the access of personnel to laser areas
- Restricting laser usage in occupied areas
- Limiting the exposure of personnel to laser radiation

These same approaches also apply to airborne systems.

In contrast to the control measures for indoor laser operation, the control of field and airborne lasers depends strongly on a determination of the hazardous zones associated with the laser application. A personnel hazard evaluation of the particular laser is needed to provide adequate guidelines on which to base the necessary control measures, including the delineation of the hazardous zones. Factors to consider in the personnel hazard evaluation include direct and reflected beam exposure and specular and diffuse reflections.

Key factors inherent in safe laser operations are the laser pointing accuracy and the extent of hazardous specular reflections.

As noted in earlier sections, many of the control measures for

Class III and Class IV lasers apply to outdoor usage in field and airborne environments. This section addresses and reemphasizes those special aspects and approaches to be considered and implemented as follows:

- Line-of-sight "weapon" approach
- Use of lasers in the airspace
- Standard operating procedures
- Laser Range Safety Officer
- Laser operator
- Radio communication
- Beam termination
- Laser reflections
- Optical instrument viewing
- Surface danger area

Line-of-Sight "Weapon" Approach

Although the laser usually lacks the ability to penetrate targets, it can be handled much like direct-fire, line-of-sight weapons, such as machine guns and rifles. Therefore, the hazard control measures used for a rifle provide most of the considerations needed to assure a safe environment for laser usage.

U.S. Army documentation on direct-fire weapon safety requirements is provided in AR 385-63.[1]

Use of Lasers in the Airspace

Among other duties, the Federal Aviation Administration (FAA) is responsible for ensuring the safety of aircraft and passengers, as well as navigable airspace.

The American National Standards Institute[2] recommends that laser experiments or programs anticipating utilization of the airspace should be coordinated with the FAA, Washington, D.C. 20590, or with any FAA regional office. This should be done in the planning stages to ensure proper control of any potential hazard to airborne personnel or equipment.

Standard Operating Procedures

Suitable procedures must be enforced for control of personnel movement in the proximity of the laser beam path, and to ensure that exposure of individuals on the ground or in flight or both is limited to established permissible levels. These operating and safety procedures are commonly referred to as SOPs or Standard Operating Procedures by the military services. As part of the SOP, a laser safety training program should be included for all personnel connected with flight and ground operations.

While safety is of paramount importance, good judgment and experience must be exercised in order not to unduly hamper the laser operation by unnecessarily stringent controls.

Laser Range Safety Officer

In order to ensure implementation of safe operating procedures and suitable administrative controls, the appointment of a single individual to serve as a Laser Range Safety Officer (LRSO) is highly desirable.

For airborne operations the officer, in conjunction with local air traffic control, will assure that strict control of traffic is maintained. Providing required information to control tower operators and ground control stations associated with the mission are other important duties. The LRSO provides adequate surveillance of the target area by ensuring that unauthorized persons do not enter, and that authorized persons use protective eyewear when so required.

Laser Operator

The laser operator has the responsibility of ensuring that firings are made only at specified targets or target areas within controlled areas and airspace. Specular surfaces should be avoided, if indeed measures have not already been taken to render them diffuse. Particular care by aircrews is required to make certain that firing is only against specified targets and target areas. Further, an airborne firing platform requires greater care by the operator, since beam wander caused by aircraft motion will be greater than that of an equivalent ground installation.

As trite as it might sound, firings should not be made if the target is not readily visible. Further, the airborne operator must ensure that the laser is properly secured and unable to fire when the aircraft is at a location other than the firing range.

Radio Communication

The use of radio communication to inform involved range personnel during the course of laser operations is essential in order to alert the personnel of the need to use protective eyewear. It is likewise necessary to keep personnel knowledgeable of the termination of laser activities, since the severe restrictions that usually accompany lasing need not be in force any longer than necessary.

Since many field operations, and particularly airborne operations, involve large land or sea areas, radio control of the operation is just about mandatory. For smaller areas, and for ground operations only, hand or flag signals may suffice. Whatever the nature of the communication approach, any break in the communication loop must be automatic cause for termination of the laser operation.

Beam Termination

Laser operations should be conducted in such a manner that the beam does not extend beyond the controlled target area. For shallow slant angles, the beam should be terminated against a natural terrain feature, such as dense trees, hills or mountains.

If the beam cannot be terminated, as would be the case for a water background bombing target, then extreme care must be exercised to restrict unprotected personnel from areas that would provide unsafe beam irradiance values.

Laser Reflections

Virtually all objects struck by a laser beam reflect some radiation, in most cases producing nonhazardous diffuse reflections. Mirror-like objects, e.g., glass, mirrors, Plexiglas, Lucite, chrome-plated metal and others, which can produce hazardous specular reflections, should be removed, covered or painted. If such objects are required or cannot with reasonable certainty be entirely removed, then protective eyewear must be worn.

Specular reflections from flat standing water, ice or snow are often considered not to provide a hazard to ground or airborne personnel not located along the azimuth of the beam path. Real-world terrain features covered with snow or ice could produce specular reflections at a variety of angles. Personnel within the direct and reflected beam path require protective eyewear. Aircrew protective visors may be required, depending on the laser output, the tactical lasing requirements and the specular hazard potential. Pilot visors must not be restrictive to the primary flying requirement. Prior to airborne lasings, a visual check for specular reflectors by the pilot or other aircrew members is most desirable. Crew members should also visually survey the target area for people, boats, and so forth, which have intruded into the area.

If the laser output is capable of producing hazardous diffuse reflections, then all objects must be removed from the beam path. The primary concern is with the immediate vicinity of the laser beam impact point, usually less than 100 feet, provided the reflections are not viewed with magnifying optical devices. For ground-based firings, it may be necessary to remove shrubs and trees near the laser.

Laser operations in rain, snow, fog or dust should be avoided unless protective eyewear is used; dew or droplets after a rainfall can also be hazardous.

Optical Instrument Viewing

Extreme care must be exercised in using optical instruments such as binoculars, telescopes, cinetheodolite viewing optics, and so on, for observation of target(s) during laser operations, since the instruments not only increase the radiation level seen by the eye but can extend the hazardous range considerably. Suitable laser filters must be placed in the optical train of the viewing instrument, with specular reflectors removed from the immediate target area and the target designed to provide diffuse reflecting properties.

Aiming the laser with the eye should be avoided to prevent looking along the axis of the beam, which increases the hazards due to reflections.

Surface Danger Area

Figure 8-4, adopted from Reference 1, is a "surface danger area" diagram for a range as would be used for a live-fire weapon. Area *B* ends at the maximum range for the ammunition. With laser firings, an additional zone, namely area *C* of Figure 8-4, may be required. Area C can be shortened or eliminated by the use of an appropriate backstop.

Aiming accuracy of the laser governs the extent of the buffer zone requirements. On the other hand, aiming accuracy of the laser device depends upon whether it is mounted on a stable platform. Recommended buffer zone requirements follow:[1]

- Two mils for stable platforms such as a reinforced bench mount, heavy duty tripod, static tank or other static base that cannot be easily moved or jarred.
- Five-mil buffer zone for a platform such as a light tripod, hand grip, moving tank or aircraft.
- Ten mils for moving platforms without gyrostabilization.

LASER PROTECTIVE EYEWEAR

After all measures have been taken to control the laser radiation hazard and the risk of accidental exposure still remains, then protective eyewear should be worn. Chapter 14 deals comprehensively with the many aspects of laser protective eyewear.

WARNING SIGNS AND LABELS

Specific requirements for warning labels have been covered for various laser classifications. This section covers the general considerations that are taken into account:

- Laser hazard symbol
- Signal words

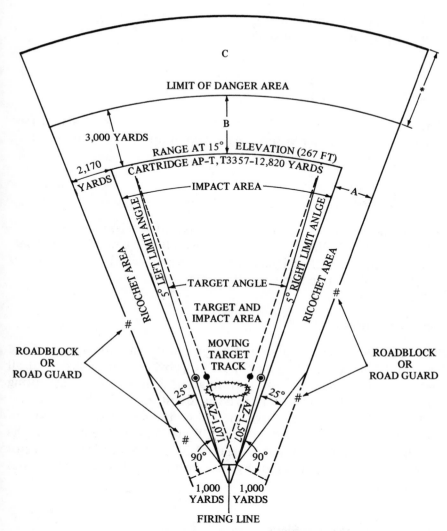

C

LIMIT OF DANGER AREA

B

3,000 YARDS

RANGE AT 15° ELEVATION (267 FT)

2,170 YARDS

CARTRIDGE AP-T, T3357–12,820 YARDS

IMPACT AREA

A

RICOCHET AREA

5° LEFT LIMIT ANGLE

TARGET ANGLE

5° RIGHT LIMIT ANLGE

RICOCHET AREA

TARGET AND
IMPACT AREA

MOVING
TARGET
TRACK

ROADBLOCK
OR
ROAD GUARD

#

ROADBLOCK
OR
ROAD GUARD

25°

AZ-1,071

AZ-1,507

25°

#

#

90°

90°

1,000
YARDS

1,000
YARDS

FIRING LINE

*Nominal hazardous range of particular laser device.

Figure 8-4. Surface danger area diagram with area *C* added. Adopted from Reference 1.

- Pertinent information
- Display of labels and signs
- Design of signs

Laser Hazard Symbol

The laser hazard symbol is comprised of a sunburst pattern (see Figure 8-1). Reference 6 specifies the color, dimensions and location of the symbol within the sign.

Signal Words

Two signal words are used with all signs and labels, namely, "CAUTION" or "DANGER." The word "CAUTION" is used for all Class II lasers and for Class IIIa lasers. The signal word "DANGER" is used for all Class IIIb and Class IV lasers.

Pertinent Information

The required signal word should be placed on the upper part of the sign or label. Adequate space must be provided for inclusion of all necessary information. This information can be included during printing of the sign or label or can be handwritten in a legible manner.

Information required above the tail of the sunburst is the type of laser, for example, pulsed ruby or CW helium-neon, etc. Below the tail of the sunburst should be special precautionary instructions or protective actions required by the reader, and radiation output information.

Display of Labels and Signs

Signs and labels should be prominently displayed in order to warn persons of potential laser radiation hazards. Use of signs and labels tends to be a more effective control measure for indoor laser installations than for outdoor installations, where a larger expanse of territory must be controlled.

Design of Signs

Reference 6 provides information needed for sign design, and includes requirements for dimension of the sign, colors, letter size, and so on. When used, laser warning signals should provide the same information as warning labels.

ALTERATION OF OUTPUT POWER OR OPERATING CHARACTERISTICS OF LASER

If intentional alterations have been made to the laser, or if during maintenance, repair or operation the output power or operating characteristics of a potentially hazardous laser have changed, then a study and/or measurements must be made to ascertain whether existing control measures are satisfactory.

REFERENCES

1. "Control of Hazards to Health from Laser Radiation," U.S. Department of the Army Technical Bulletin TB-MED-279, Washington, D.C., May 1975.
2. "American National Standard for the Safe Use of Lasers, Z136.1-1976," American National Standards Institute, New York, 1976.
3. "Laser Products Peformance Standards," *Federal Register*, Vol. 40 No. 148, pp. 32252–32266, U.S. Dept. of Health, Education and Welfare, Food and Drug Administration, Bureau of Radiological Health (BRH), Washington, D.C., 31 July 1975.
4. New York State Industrial Code Rule 50, "Lasers," State of New York Department of Labor, New York, 1 Aug. 1972.
5. Sliney, D. H., Vorpahl, K. W., Winburn, D. C., "Environmental Health Hazards From High-Powered, Infrared Laser Devices," *Archives of Environmental Health*, Vol. 30, American Medical Association, pp. 174–179, April 1975.
6. "American National Standard Specifications For Accident Prevention Signs, Z35.1," American National Standards Institute, New York, 1972.

BIBLIOGRAPHY

1. Van Pelt, W. F., Stewart, H. F., Peterson, R. W., Roberts, A. M., Worst, J. K., *Laser Fundamentals and Experiments*, U.S. Department of Health, Education, and Welfare, Public Health Service, Bureau of Radiological Health,

Southwestern Radiological Health Laboratory, Las Vegas, Nevada, May 1970.

2. *Laser Safety Guide*, Third Edition, Laser Institute of America, Cincinnati, Ohio, Jan. 1976.

3. Chabot, L., Mallow, A., Wengler, N., "Airborne Testing of Advanced Multisensor Aircraft," Fifth Annual Symposium Proceedings, Society of Flight Test Engineers, pp. 2-95 to 2-128, Aug. 1974.

9

Control of Associated Laser Hazards

Each of the associated laser hazards cited in Chapter 4 is addressed here on an individual basis. Reduction of the associated hazards will evolve through evaluation of the potential hazards and the taking of appropriate control measures so that operating personnel are provided a safe work environment. The nature and purpose of this handbook precludes in-depth treatments of the many diverse disciplines dealt with. However, numerous references are cited to acquaint the reader fully with the many regulations, procedures and miscellaneous information needed for properly controlling laser-associated hazards. A discussion is included that details first aid procedures to be used for serious electrical shock victims.

ELECTRICAL HAZARD CONTROLS

Potential electric shocks associated with laser equipment can be quite serious and even fatal. This section is divided into the subsections listed below, with the goal of controlling the possibility of sustaining hazardous electrical shocks. Since no control systems will ever be perfect, a discussion of first aid for severe electric shock victims is provided.

- Regulations, procedures and standards
- General safety considerations for preventing electric shock
- Design and user precautions for high-powered laser systems
- First aid for severe electrical shock victims

Regulations, Procedures and Standards

Many regulations and procedures exist that deal with the electrical hazard. The Chemical Rubber Company's *Handbook of Laboratory Safety* has excellent sections dealing with electrical safety[1] and grounding electrical equipment.[2] Franks and Sliney[3] consider the Atomic Energy Commission's "Electrical Safety Guides for Research Safety and Fire Protection"[4] a very useful publication, and it is also given as a reference in ANSI's "Standard for the Safe Use of Lasers, Z136.1-1976."[5] Safety in general and the need to ground electrical chassis are considered in MIL Standard 454 C, although high voltage power supplies are just lightly addressed.[3]

As for electrical installation of laser equipment, ANSI[5] recommends that it be done in accordance with the "American National Standard National Electrical Code, C1-1971 (NFPA 70-1971), Articles 300 and 400." Equipment installed in this manner is acceptable to the Occupational Safety and Health Administration (OSHA) as long as it is determined safe by a nationally organized testing laboratory such as Underwriters Laboratories, Inc. or Factory Mutual Corporation.

General Safety Considerations For Preventing Electric Shock

Table 9-1 summarizes a number of key general safety guidelines to be used for prevention of electrical shock in working with electrical equipment.

Design and User Precautions for High-Powered Laser Systems

Certain precautions taken by the designer and the user of high-powered laser systems can appreciably reduce the probability that personnel will obtain serious electrical shocks. Table 9-2 summarizes many such precautions.

**Table 9-1. Summary of General Safety Consideration For
Preventing Electrical Shock.**

1. All electrical equipment should be treated as if it were "live."
2. A "buddy system" should be used for work on live electrical equipment, and is often mandated by many organizations, particularly after normal working hours or in isolated areas. Ideally the persons should be knowledgeable of first aid and cardiopulmonary resuscitation.
3. Rings and metallic watchbands should not be worn, nor should metallic pens, pencils or rulers be used, while one is working with electrical equipment.
4. Live circuits should be worked on using one hand, when it is possible to do so.
5. When one is working with electrical equipment, only tools with insulated handles should be used.
6. Immediately report for repair or investigation any electrical equipment that upon touch gives the slightest perception of current.
7. When working with high voltages, one should consider floors conductive and grounded unless he is standing on a suitably insulated dry matting normally used for electrical work.
8. Live electrical equipment should not be worked on when one is standing on a wet floor, or when the hands, feet or body is wet or perspiring.
9. One should not undertake hazardous activities when truly fatigued, or under the influence of medication that dulls or slows the mental and reflex processes, or with a mental attitude that would incline him toward risk-taking, be it via emotional or chemical stimulus.

First Aid for Severe Electrical Shock Victims

Severe electrical shock can cause both termination of breathing and stoppage of the heartbeat. First aid or emergency cardiopulmonary resuscitation requires the following key steps:

- Opening of the airway
- Restoration of breathing
- Restoration of heartbeat or circulation

after one has opened the electrical circuit, or removed the victim still in contact with the circuit using an insulated object. Professional assistance should be sought as quickly as possible. The possibility of breathing restoration is good if measures are taken within two minutes but becomes poor after a delay of about four minutes, although some individuals have been revived after delays of half an hour or more.[6]

Table 9-3 summarizes the essential first aid procedures to be used in assisting victims of severe electrical shock.

**Table 9-2. Summary of Design and User Precautions for
High-Powered Laser Systems.**

1. Circuits and components with peak open circuit potentials greater than 42.5 volts should be considered a shock hazard, unless the maximum current possible is less than 0.5 milliampere.[5] Positive protection is required against accidental contact, such as coverings or enclosures.
2. Some high voltage hazard points to be wary of follow:

    • Capacitor bank terminals
    • Transformer terminals
    • Cathode ray tube terminals
    • Circuit or meter adjusting screws
    • Fuse panels

3. Interlocks should be used to remove voltages from accessible parts when doors or panels are opened for servicing of equipment.
4. All laboratory electrical services should be grounded and tested to ensure that a suitable and continuous grounding system exists.
5. All chassis and exposed metal parts of laser equipment should be grounded. Reliable and continuous bonding between the item to be grounded and the grounding conductor of the power wiring system is essential.
6. Bleeder resistors should be used to handle capacitor discharge currents. However, work on the capacitor(s) should not be attempted unless the charge has been safely dissipated. (See item 8 below.)
7. Large capacitors can be an explosion hazard if subjected to voltages exceeding their rating, and may therefore require a secured location. The capacitors should be regularly checked for signs of deformation or leaks.
8. Dissipation of the charge on high voltage capacitors must be completed prior to servicing of interlocked enclosed equipment. A solid metal grounding rod with an insulating handle, and capable of handling the energy, must be well secured to ground (spring clip may be inadequate) prior to making contact with the capacitor contact which is above ground. In cases where neither side is grounded, a safe shorting bar arrangement should be used.
9. Crowbars, grounding switches and other safety devices should be designed and installed in such a manner as to handle the mechanical forces present when faults occur or crowbar currents flow.
10. Fault-current-limiting devices such as fuses, circuit breakers or resistors should be able to dissipate the total energy. Over-capacity fuses should not be used to mask defective circuits.
11. Safety apparatus, such as face shields, safety glasses, insulating mats, rubber gloves and others, should be supplied and used as needed.
12. Appropriate precautionary information should be made available to personnel using electrically hazardous lasers.
13. The user should make sure that each laser system is permanently marked with its primary electrical rating in volts, frequency, watts or amperes.

Table 9-3. Summary of First Aid Procedures for Victims of Severe Electrical Shock.

Procedures if Breathing Has Ceased	Procedures if Heart Has Stopped Beating	General Procedures
1. OPEN THE AIRWAY IMMEDIATELY–place victim on his back, remove foreign objects from the mouth or throat and with one hand lift the neck and with the other tilt the head backwards. This keeps the tongue out of the airway. If breathing is restored by the head tilt, check for heartbeat (step 5).	6. BEGIN EXTERNAL CARDIAC COMPRESSION–per steps 7–11.	• CALL FOR HELP IMMEDIATELY–if possible.
2. START RESCUE BREATHING–if victim is not breathing or breathing is feeble. Pinch victim's nostrils, open your mouth wide making a tight seal about victim's mouth and blow hard, unless victim is infant or young child. Remove mouth to allow victim to exhale. Alternate method is to close victim's mouth and blow through his nose.	7. The victim should be on a firm surface, but compression should not be delayed if this is a problem.	• Continue mouth-to-mouth resuscitation and/or heart massage while the victim is being taken to the hospital, or until he is revived, or until told by the physician to stop.
3. DETERMINE IF OBSTRUCTION EXISTS–the first blowing tells if an obstruction exists, if the chest does not rise. If obstruction exists, place victim on side and apply hard blows to the back between the shoulder blades.	8. Put the lower palm of one hand on the lower half of the victim's breastbone, with the other hand on top of the first.	
4. REPEAT RESCUE BREATHING–cycle is repeated every 4–5 seconds.	9. Keep your arms straight and push downward from the shoulders with sufficient force to move the victim's chest 1½ to 2 inches.	
5. DETERMINE IF HEART IS BEATING–check carotid artery pulse after 3–5 breathing cycles, by placing two fingers about the sides of the Adam's apple. If no pulse is felt, start external cardiac compression (step 6). If heart is beating, continue rescue breathing until victim's normal breathing is restored or until a physician tells you to stop.	10. Repeat at a rate of about one per second.	
	11. Alternate between heart massage and mouth-to-mouth or mouth-to-nose breathing.	
	• If alone with the victim, perform two quick breaths and then 15 heart compressions.	
	• If you have help, the ratio is one breath to five heart compressions.	

CONTROL OF AIRBORNE CONTAMINANTS

User personnel exposed to hazardous airborne contaminants associated with laser operations should be safeguarded by keeping concentrations within established safe limits, which is done by resorting to the following control techniques:

- Ventilation
- Contaminant detection instrumentation
- Respiratory protective devices

Ventilation Control

Ventilation control can be broadly approached from two standpoints—outdoor and indoor requirements.

The inherent dilution of airborne contaminants outdoors during laser usage will, in general, limit concentrations to safe levels because of air movement and the large air volume available for dilution. Thus, this natural form of dilution ventilation is a frequently used and proven control technique.

Indoor ventilation control techniques are obviously more difficult and costly. Spot ventilation, which exhausts airborne contaminants near their point of origin, is an excellent control for preventing inhalation of hazardous contaminants. Local exhaust hoods and fans for ventilation are other approaches. Any of the approaches must be based upon a knowledge of the process or operation and contaminant toxicity for which control must be obtained.

Additional material related to ventilation and exhaust systems can be found in References 7 and 8.

Contaminant Detection Instrumentation

For large-type laser systems, particularly those where gas or liquid bulk storage facilities exist, provision should be made for contaminant detection instrumentation. This instrumentation can assure maintenance of contaminant levels under allowable exposure limits, and enable early detection of system leaks.

Respiratory Protective Devices

Respiratory protective devices should be available for use in coping with emergencies. Emergency equipment should include approved air-supplied breathing devices. The area selected for storage of the respiratory devices should be one that permits placement in a contaminant-free atmosphere.

Compressed Gases

The following regulations, standards and guides should be observed while using compressed gases:[5]

- "American National Standard Method for Marking Portable Compressed Gas Containers to Identify the Material Contained, Z 48.1-1971 (R-1954)"
- "American National Standard Compressed Gas Cylinder Valve Outlet and Inlet Connections, B57.1-1965"
- *Handbook of Compressed Gases*, Compressed Gas Association, New York, 1966
- Pertinent regulation of the U.S. Department of Transportation

CONTROL OF CRYOGENIC LIQUIDS HAZARD

ANSI[5] recommends that cryogenic liquids be stored and handled in accordance with instructions provided in the *Handbook of Compressed Gases*, prepared by the Compressed Gas Association in 1966.

A survey treatment of some controls, precautions and actions to be observed in dealing with cryogenic liquids follows:

- Personnel should avoid wearing garments that can catch (e.g., pocket and trouser cuffs) cryogenic fluids near the flesh.
- Jewelry should not be worn when working with cryogenics.
- If contact with cryogenics is possible, full-face protection is needed, as are proper outer garments, insulated gloves and high shoes or boots.
- If cryogenics spill on the skin, the area should be flooded with a large amount of unheated water and then cold compresses applied.

- If toxic cryogenics are used, respiratory equipment should be available.
- Equipment should be kept very clean and contaminants avoided that can create hazardous situations when mixed with cryogenics, particularly liquid or gaseous oxygen.
- Work areas with cryogenics should be monitored to automatically warn personnel at the outset of dangerous conditions.
- Dumping "inert" liquid nitrogen cryogenic requires sufficient ventilation. Otherwise exclusion of oxygen from the lungs of personnel in the area can cause unconsciousness or death.

CONTROL OF NOISE HAZARD

Hearing-conservation measures, which include the use of protective devices, should be implemented when an individual is exposed to steady noise levels above 90 dB for eight hours a day. Exposures up to 115 dB are, however, permitted for periods under 15 minutes, while the maximum allowed exposure is 140 dB peak sound pressure levels.[9]

CONTROL OF IONIZING RADIATION HAZARD

Adequate shielding must be employed for use with high voltage power supply tubes with potentials exceeding 15 KV, since X-rays can be generated. ANSI[5] states that the source, quantity and intensity of X-rays emanating from laser power supplies and components should be investigated and controlled in accordance with provisions listed in the "American National Standard Safety Standard for Non-Medical X-ray and Sealed Gamma-Ray Sources, Z54, 1-1963" (*NBS Handbook H93*). Applicable federal, state or local codes and regulations must also be complied with.

CONTROL OF NON-LASER BEAM OPTICAL RADIATION HAZARD

Ultraviolet radiation emanating from laser discharge tubes and pumping lamps (not from the primary laser beam) has to be suitably shielded to prevent a potentially hazardous situation. Certain plastics and heat-resistant glasses readily attenuate ultraviolet radiation. Reference 10 provides additional information on ultraviolet radiation hazards.

CONTROL OF EXPLOSION HAZARD

High-pressure arc lamps and filament lamps in laser equipment must be enclosed in housings capable of withstanding the maximum explosive pressures resulting from lamp disintegration. The laser target and elements of the optical train that may shatter during laser operation must also be enclosed or similarly protected to prevent injury to operators and observers.

Since electrolytic capacitors can explode if voltages exceed their rating, such capacitors should be tested to make sure they can resist the highest probable potentials should other circuit components fail. If the capacitors are adequately housed and no potential hazard to personnel exists, these tests need not be done.

CONTROL OF FIRE HAZARD

ANSI[5] recommends that components made of combustible material, such as transformers, that do not pass a short-circuit test without ignition be provided with individual noncombustible enclosures. For further information refer to the American National Standard Safety Standard for Radio and Television Receiving Appliances, C33.55-1969 (UL 492—June 1969). Power supply circuit wiring must be completely enclosed in noncombustible material. Combustible materials should not be kept near laser equipment.

Some modest fire-fighting equipment like estinguishers should be placed in the proximity of laser equipments. The purpose is control and estinguishing of small fires only.

REFERENCES

1. Ehrenkranz, T. E., Marsischky, G. W., "Electrical Equipment, Wiring and Safety Procedures," *CRC Handbook of Laboratory Safety*, N. V. Steere, Editor, 2nd Edition, Chemical Rubber Co., Cleveland, Ohio, pp. 528–538, 1971.
2. Electronic Industries Association, "Grounding Electronic Equipment,"

CRC Handbook of Laboratory Safety, N. V. Steere, Editor, 2nd Edition, Chemical Rubber Co., Cleveland, Ohio, pp. 516–519, 1971.

3. Franks, J. F., Sliney, D. H., "Electrical Hazards of Lasers," *Electro-Optical Design*, pp. 20–23, Dec. 1975.
4. "Electrical Safety Guides for Research, Safety and Fire Protection," U.S. Atomic Energy Commission Technical Bulletin No. 13 (0-292-830), Government Printing Office, Washington, D.C., 1968.
5. "American National Standard for the Safe Use of Lasers, Z136.1-1976," American National Standards Institute, New York, 1976.
6. Price, L. D., "Codes and Standard Practices—Resuscitation," *Standard Handbook for Electrical Engineers*, D. G. Fink, Editor, 10th Edition, Section 29, 1971.
7. Steere, N. V., "Ventilation of Laboratory Operations," *CRC Handbook of Laboratory Safety*, N. V. Steere, Editor, 2nd Edition, Chemical Rubber Co., Cleveland, Ohio, pp. 141–149, 1971.
8. American Society of Heating, Refrigerating and Air-Conditioning Engineers, Inc., "Exhaust Systems," *CRC Handbook of Laboratory Safety*, N. V. Steere, Ed., 2nd Edition, Chemical Rubber Co., Cleveland, Ohio, pp. 150–153, 1971.
9. "Federal Register, Safety and Health Regulations for Construction," Vol. 36, No. 75, Part II, Section 1518.52, Bureau of Labor Standards, Dept. of Labor, 17 April 1971.
10. Sliney, D. H., Freasier, B. C., "The Evaluation of Optical Radiation Hazards," *Applied Optics*, Vol. 12, No. 1, pp. 1–22, 1973.

BIBLIOGRAPHY

1. "Cardiopulmonary Resuscitation," *CRC Handbook of Laboratory Safety*, N. V. Steere, Editor, 2nd Edition, Chemical Rubber Co., Cleveland, Ohio, pp. 25–35, 1971.
2. Cutler, A., "How To Use Resuscitative Techniques," *Four Minutes to Life*, Cowles Book Co., Inc., New York, pp. 149–177, 1970.
3. *Emergency Family First Aid Guide*, Simon & Schuster, New York, 1970.
4. Handley, W., Editor, *Industrial Safety Handbook*, McGraw-Hill, London, 1969.
5. Carrier, W. H., Cherne, R. E., Grant, W. A., Roberts, W. H., *Modern Air Conditioning, Heating, and Ventilating*, Pitman Publishing Corp., New York, 1959.
6. *Laser Safety Guide*, Third Edition, Laser Institute of America, Cincinnati, Ohio, Jan. 1976.
7. Spencer, E. W., "Cryogenic Safety," *CRC Handbook of Laboratory Safety*, N. V. Steere, Editor, 2nd Edition, Chemical Rubber Co., Cleveland, Ohio, pp. 25–35, 1971.

10

Public Laws

INTRODUCTION

Government laser safety regulations (standards) carry the greatest weight, since they can be legally enforced, as contrasted to voluntary standards such as those developed and published by ANSI[1] and ACGIH.[2] Federal regulations are in existence for laser manufacturers and users of construction lasers, and are in preparation for broad user applications. In addition, standards can and have been enacted by state agencies and foreign countries. International organizations likewise play a role in influencing the safe use of lasers, e.g., the World Health Organization (WHO) and the International Electrotechnical Commission (IEC).

The emphasis in this chapter will be on federal and state regulations or standards.

On the federal side, the congressional acts authorizing the various departmental agencies to implement laser safety standards will be addressed. The departmental agency standards themselves, along with documentation requirements and guides, are discussed. Further, a comprehensive tabular summary is included relating the requirements of the "Laser Products Performance Standards"[3] to the enforcing agency's interpretation of those standards.

A survey of existing and proposed state legislation is then covered. Particular emphasis is given to New York State Industrial Code Rule 50,[4] governing the safe use of lasers in that state. A comprehensive interpretation of Code Rule 50 is provided, with a copy of that regulation included as Appendix B.

FEDERAL LASER SAFETY LEGISLATION

Congressional acts have authorized two executive branch departments to develop federal standards for lasers. Figure 10-1 summarizes the congressional acts, responsible executive departments and agencies and specific regulations and agency documentation guides for compliance with these regulations.

The "Radiation Control for Health and Safety Act of 1968 (Public Law 90-602)" gives the U.S. Department of Health, Education and Welfare jurisdiction to develop product performance standards that apply to laser products manufactured for sale. Specific responsibility within that department rests with the Bureau of Radiological Health (BRH). From Figure 10-1, it is seen that the BRH was the agency which prepared the "Laser Products Performance Standards,"[3] dated July 31, 1975. The effective date for compliance by manufacturers is August 2, 1976. Enforcement of compliance rests with the BRH. In order to assist manufacturers in the preparation of initial and model change reports and to comply with the performance standard set forth in Sections 1040.10 and 1040.11, and permit certification as set forth in Section 1010.2, the BRH has prepared a "Guide For Submission of Information on Lasers and Products Containing Lasers Pursuant to CFR 1002.10 and 1002.12."[5] In addition, "Quality Control Practices for Compliance with the Federal Laser Product Performance Standard"[6] was prepared to assist manufacturers in developing and implementing quality control and testing programs so as to ensure compliance with the standard, and is intended for use in conjunction with the Reference 5 report.

The "Occupational Health and Safety Act of 1970 (Public Law 91-596)" gives the U.S. Department of Labor jurisdiction to develop laser user standards. Responsibility within that department rests with two agencies, as can be seen from Figure 10-1. The Occupational Safety and Health Administration (OSHA) is in the process of developing broad user standards. The Bureau of Labor Standards

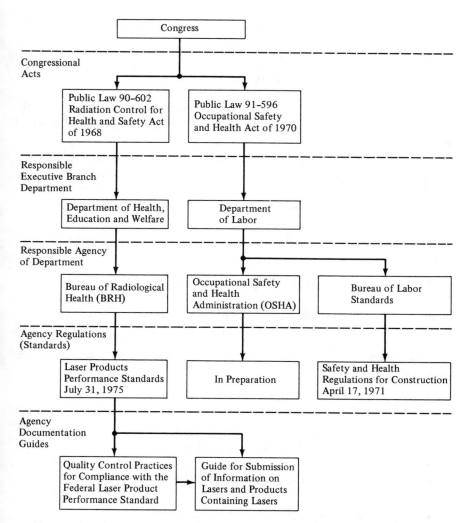

Figure 10-1. Summary flow of federal laser safety regulations (standards).

has prepared safety standards for laser usage in construction in "Safety and Health Regulations for Construction."[7]

The above discussion has highlighted two major categories regarding laser safety standards, namely,

- Product performance standards for laser manufacturers
- User safety standards

What follows addresses numerous aspects of the two categories of standards.

Product Performance Standards For Laser Manufacturers

The requirements placed upon laser manufacturers for achieving compliance and certification of their product in accordance with the Laser Products Performance Standard are difficult to implement. Further, interpretation of that standard requires very careful use of BRH's report guide of Reference 5 and the quality control practices guide of Reference 6. A manufacturer is obliged to submit a Laser Product Report (in accordance with Reference 5) to the BRH for each new or changed model, prior to introduction of the product into commerce. The written documentation serves two main purposes:

- To assure the BRH that the product complies with all provisions of the standard.
- To demonstrate to the BRH that the associated testing program and the results of such program are adequate.

The Laser Product Report covers the following ten areas in considerable detail:

- Manufacturer and product identification
- Laser product and model identification
- Compliance with the labeling requirements
- Compliance with the informational requirements
- Description of the laser product
- Reporting levels of accessible radiation and classification of the laser product
- Compliance with performance requirements
- Quality control tests and testing procedures for compliance
- Life and endurance testing
- Instrumentation and calibration

Table 10-1 provides a correlation between individual requirements of the Laser Product Performance Standard and the multitude of items covered in the ten areas cited above.

Appendix C, Class Level of Accessible Laser Radiation, has been

Table 10-1. Summary of BRH Laser Product Performance Standards Requirements.

Requirements Per Laser Products Performance Standards (Part 1040)			BRH Report Requirements**	
Section 1040.10 Para. No.	General Description	Part & Para. No.	Laser Class	Description
1040.10	LASER PRODUCTS	—	All	INTRODUCTION–GENERAL INSTRUCTIONS FOR THE LASER PRODUCT REPORT
	(a) Applicability Provisions of this section and Section 1040.11 (Specific Purpose Laser Products) are applicable to all laser products manufactured or assembled on or after August 2, 1976. Exceptions follow:	(a)		Purpose • This guide is intended to assist manufacturers in preparation of initial and model change reports required by 21 CFR 1002.10 and 1002.12 for LASER PRODUCTS that are to comply with the performance standard set forth in Sections 1040.10 and 1040.11, and that are to be certified by the manufacturer as set forth in Section 1010.2. • The manufacturer is to provide BRH, through written documentation, with information concerning the manufacturing and distribution of the product: —To assure that product complies with all provisions of the standard. —To demonstrate that the associated testing program and the results of such program are adequate. General
	• Laser product sold to manufacturer of electronic product for use as a component (or replacement), or	(a) (1)		• This report must be submitted to the BRH prior to the introduction of the certified product into commerce. • The performance standard applicable to laser products is for product performance rather than product design, and provides product-related safety constraints. • Report completely on each model of laser product. • If particular request (general point or specific
	• Laser product sold by or for a manufacturer of electronic product for use as a component (or replacement) provided that	(a) (2)		

Table 10-1. (Continued)

Requirements Per Laser Products Performance Standards (Part 1040)		BRH Report Requirements**		
Section 1040.10 Para. No.	General Description	Part & Para. No.	Laser Class	Description
	such laser product:			question) for information is not appropriate to laser product, declare positively that this is the case and explain why this is so.
(a) (2) (i)	–Is accompanied by adequate instruction for installations in electronic product.			• In general, any aspect of the product pertaining to radiation safety should be reported, including those aspects that are not covered by the guide and those for which there are no applicable provisions as set forth in Sections 1040.10 and 1040.11.
(a) (2) (ii)	–Is labeled with statement that it is designed for use solely as a component of such electronic product and therefore does not comply with the appropriate requirements of this section and section 1040.11 for complete laser tests, and			• All answers should be clear, concise and complete.
(a) (2) (iii)	–Is not a removable laser system as described in paragraph (c) (2) of this section,			
—	—	1	—	MANUFACTURER AND REPORT IDENTICATION
—	—	1.1	—	Manufacturer to provide:

		• Address
		• Corresponding official's name, title, telephone number and signature
	1.2	Importer (if applicable) to provide:
		• Name
		• Address
		• Corresponding official's name, title, telephone number and signature
	1.3	State report type (e.g., initial, model change or supplement to report) and date submitted.
	1.4	Provide report date.
	2.	LASER PRODUCT AND MODEL IDENTIFICATION
	2.1	List name, model number and model family designation of laser product being reported.
	2.2	Report any other brand name, model number and name and address of each company under whose name the reported model is sold.
	2.3	State owner of reported product design and/or authorized company official who can make final decisions relating to design, manufacture and marketing of product.
	2.4	State if report is submitted pursuant to 21 CFR 1002.61 (c) of the Regulations for the Administration and Enforcement of the Radiation Control for Health and Safety Act of 1968.
1040.10 Laser Products	2.5	State if the standards for certification of the reported model are those of the performance standard for laser products (21 CFR 1040.10 and 1040.11).
1040.11 Specific Purpose Laser Products	2.6	State the advertised, intended and known functions (uses) of the reported model. If general-purpose laser, so state.
	2.7	State if product or part of product is intended by design to be used as a component in a product manufactured by another manufacturer.

Table 10-1. (*Continued*)

| Requirements Per Laser Products Performance Standards (Part 1040) | | Part & Para. No. | Laser Class | BRH Report Requirements** |
Section 1040.10 Para. No.	General Description			Description
				If yes:
—	—	2.7.1	—	• Provide name and address of manufacturer, where known.
—	—	2.7.2	—	• Identify the products (including model numbers) which use the reported product as a component, where known.
(i)	Modification of a Certified Product	2.8	—	State if the reported laser product is the result of the modification of a certified product [Section 1040.10 (i)]. If yes, then report all information on the prior certification and identification labels.
1010.2	Certification	2.9	—	State if reported laser product model has a label certifying that the product conforms to the provisions of 21 CFR 1040.10 and 1040.11, as required by 21 CFR 1010.2.
1010.3	Identification	2.10	—	State if the reported laser product has a label that conforms to the provisions of 21 CFR 1010.3.
1010.3	Identification	2.11	—	State if laser product incorporates a certified laser product. If yes, then report the information required to be on the identification label of that certified laser product pursuant to Section 1010.3.
(c) (2)	Removable Laser System • Any laser system which is incorporated into a laser product and which is capable, without modification, of producing laser radiation when removed	2.12	—	State if product incorporates a removable laser system as defined in Section 1040.10 (c) (2). If yes, then identify the removable laser system by manufacturer and product model number.

from such laser product, shall itself be considered a laser product, and shall be separately subject to the applicable requirements in this subchapter for laser products of its class.	—	2.13	State if the laser product as introduced into commerce is supplied with a laser or laser system. If no,
	—	2.13.1	• Report which laser or laser system, by manufacturer and model number, is recommended for incorporation into the product.
	—	2.13.2	• If you do not recommend a specific laser or laser system for incorporation into the reported product, state the specifications of the laser or laser system to be incorporated.
(b) (22) Medical Laser Product • Any laser product manufactured, designed, intended or promoted for purposes of *in vivo* diagnostic, surgical or therapeutic laser irradiation of any part of the human body.	—	2.14	State if laser product is a medical laser product as defined in Section 1040.10 (b) (22). If yes, then state class of laser (i.e., I, II, III or IV). *Note:* If you confirm that product is a medical laser product, then be sure to include sufficient description in Parts 3.4 and 7.14 to assure the BRH of compliance with Section 1040.11 (a) (medical laser products requirements).
b(35) Surveying, Leveling, or Alignment Laser Product • Laser product manufactured, designed,	—	2.15	State if laser product is a surveying, leveling or alignment laser product as defined in Section 1040.10 (b) (35). *Note:* If you confirm that your product is a surveying, leveling or alignment laser product, then be

Table 10-1. (Continued)

| Requirements Per Laser Products Performance Standards (Part 1040) | | | BRH Report Requirements** |
Section 1040.10 Para. No.	General Description	Part & Para. No.	Laser Class	Description
	intended or promoted for one or more of the following uses: (i) Making angular measurements. (ii) Positioning or adjusting parts relative to one another. (iii) Defining a plane, level, elevation or straight line.			sure to include in Part 6 sufficient description to assure the BRH of compliance with Section 1040.11 (b) (surveying, leveling and alignment products requirements).
b(10)	Demonstration Laser Product • Any laser product manufactured, designed, intended or promoted for purposes of demonstration, entertainment, advertising display or artistic composition.	2.16	–	State if laser product is a demonstration laser product as defined in Section 1040.10 (b) (10). Note: If you confirm that your laser product is a demonstration laser product, then be sure to include in part 6 sufficient description to assure the BRH of compliance with Section 1040.11 (c) (demonstration laser products requirements).
–	LABELING	3	–	COMPLIANCE WITH THE LABELING REQUIREMENTS

1010.2	Certification (Label)	3.1 3.1.1 3.1.2 3.1.3	All	**Certification Label** • Submit sample of certification label, or facsimile if label not available at time of reporting. • Label should be legible and clearly visible when reported model is fully assembled and ready for use. • Certification label must be so positioned as to make unnecessary during reading, human exposure to laser or collateral radiation in excess of accessible emission limits of Class I and Table 7-1A.
(g) (9)	Positioning of Labels	3.1.4		• State if label is permanently affixed or inscribed on the reported product. *Note:* Certification label must be identified and its location indicated in Part 5.1.
1010.3	Identification (Label)	3.2 3.2.1 3.2.2 3.2.3 3.2.4	All	**Identification Label** • Submit sample of identification label, or facsimile if label not available at time of reporting. • If place of manufacture as stated on label is coded, provide that code. • State if month and year of manufacture is provided clearly and legibly without abbreviation, and with year shown as four-digit number. For example, Manufactured: August 1976 • Label should be legible and clearly visible when reported model is fully assembled and ready for use.
(g) (9)	Positioning of Labels	3.2.5		• Identification label must be so positioned as to make unnecessary during reading, human exposure to laser or collateral radiation in excess of accessible emission limits of Class I and Table 7-1A.
		3.2.6		• State if label is permanently affixed or inscribed on the reported product. *Note:* Identification label must be identified and its location indicated in Part 5.1.

Table 10-1. *(Continued)*

Requirements Per Laser Products Performance Standards (Part 1040)		BRH Report Requirements**		
Section 1040.10 Para. No.	General Description	Part & Para. No.	Laser Class	Description
(g) (1)	Class II Designation and Warning	3.3	II, III, IV	Warning Logotypes
(g) (2)	Class III Designation and Warning			
(g) (3)	Class IV Designation and Warning			
(g) (5)	Radiation Output Information			
(g) (8)	Warning For Invisible Radiation			
(g) (9)	Postioning of Labels			
(g) (10)	Label Specifications			
		3.3.1		• Submit sample of warning logotype, or facsimile if label not available at time of reporting.
		3.3.2		• State if warning logotype is legible and clearly visible when reported model is fully assembled and ready to use.
		3.3.3		• Warning logotype must be so positioned as to make unnecessary during reading, human exposure to laser or collateral radiation in excess of accessible emission limits of Class I and Table 7-1A.
		3.3.4		• State if warning logotype is permanently affixed or inscribed on the reported product.

Notes: (1) Logotype must be adequately identified and its location indicated in Part 5.1.

(2) Adequate information must be included in Part 6.1 to determine compliance of product with Section 1040.10 (g) (5) (radiation output information).

(3) Adequate information must be included in Part 6.1 to determine compliance of the product with Section 1040.10 (g) (8) (warning for invisible radiation).

3.4 II, III, IV Aperture Label (nonmedical and medical laser products)

3.4.1
• For each aperture identified in Part 5 of Class II, III or IV nonmedical and medical laser products, submit a sample of required aperture label, or facsimile if label is not available at time of reporting.

3.4.2
• State if aperture label is legible and clearly visible when reported model is fully assembled and ready for use.

3.4.3
• Aperture label must be so positioned as to make unnecessary during reading, human exposure to laser or collateral radiation in excess of accessible emission limits of Class I and Table 7-1A.

3.4.4
• State if aperture label is permanently affixed or inscribed on reported product.
Notes: (1) Aperture label must be adequately identified and its location indicated in Part 5.
(2) For nonmedical laser products, adequate

(g) (4) Aperture Label (non-medical)

(g) (8) • Warning For Invisible Radiation

(g) (9) • Positioning of Labels
(g) (10) • Label Specifications
1040.11 Medical Laser Products
(a) (3) (Aperture Label)

Table 10-1. (*Continued*)

| Requirements Per Laser Products Performance Standards (Part 1040) | | | BRH Report Requirements** |
Section 1040.10 Para. No.	General Description	Part & Para. No.	Laser Class	Description
				information must be included in Part 6.1 to determine compliance of product with Section 1040.10 (g) (8) (warning for invisible radiation).
(g) (6)	Labels For Noninterlocked Protective Housings	3.5	All	Labels For Noninterlocked Protective Housings
(g) (8)	• Warning For Invisible Radiation			
(g) (9)	• Positioning of Labels	3.5.1		• For each portion of protective housing identified in Part 5 which has no safety interlock and which is designed to be displaced or removed during operation, maintenance or service, and thereby could permit human access to laser or collateral radiation in excess of limits of Class I and Table 7-1A, submit a sample of the required label, or facsimile if label is not available at time of reporting.
(g) (10)	• Label Specificiation			
		3.5.2		• State if label of Part 3.5.1 is legible and clearly visible on protective housing prior to displacement or removal of such portion of protective housing and clearly visible on product in close proximity to opening created by removal or displacement of such portion of protective housing.
		3.5.3		• State if label of Part 3.5.1 is so positioned as to make unnecessary during reading, human exposure to laser or collateral radiation in excess of accessible emission limits of Class I and Table 7-1A

3.5.4   • State if label of Part 3.5.1 is permanently affixed or inscribed on reported product.

Notes: (1) There are four possible label selections for the warning pertaining to accessible laser radiation (Class II, IIIa, IIIb and IV), and two selections for the warning pertaining to collateral radiation (nonlaser radiation for λ = 250–13,000 nm and X-radiation).

(2) Adequate identification and description of non-interlocked portion of protective housing must be included in Part 5.

(3) Label(s) must be adequately identified and location(s) indicated in Part 5.

(4) Adequately identify in Part 5 the accessible laser and collateral radiation associated with the non-interlocked portion of the protective housing, and adequately describe in Part 6 the radiation characteristics of the laser and collateral radiation made accessible by virtue of the noninterlocked portion of the protective housing.

3.6   All   Labels For Defeatably Interlocked Protective Housing

(g) (7)   Labels For Defeatably Interlocked Protective Housing
(g) (8)   • Warning For Invisible Radiation
(g) (9)   • Positioning of Labels
(g) (10)   • Label Specification

3.61   • For each defeatably interlocked portion of the protective housing identified in Part 5 which is designed to be displaced or removed during operation, maintenance or service, and which upon interlock defeat could permit human access to laser or collateral radiation in excess of the limits of Class I or Table 7-1A, submit a sample of

Table 10-1. (*Continued*)

Requirements Per Laser Products Performance Standards (Part 1040)		Part & Para. No.	Laser Class	BRH Report Requirements**
Section 1040.10 Para. No.	General Description			Description
		3.6.2		required label, or facsimile if the label is not available at the time of reporting. • State if label of Part 3.6.1 is legible and clearly visible on product prior to and during interlock defeat and in close proximity to opening created by removal or displacement of such portion of protective housing.
		3.6.3		• State if label of Part 3.6.1 is so positioned as to make unnecessary during reading, human exposure to laser or collateral radiation in excess of the accessible emission limits of Class I and Table 7-1A.
		3.6.4		• State if label of Part 3.6.1 is permanently affixed or inscribed on the reported product. *Note:* Notes (1) through (4) of Part 3.5 above apply to this Part for defeatably interlocked protective housing.
(h)	INFORMATIONAL REQUIREMENTS	4	—	COMPLIANCE WITH THE INFORMATIONAL REQUIREMENTS
(h) (1)	User Information	4.1	All	User Information
(h) (1) (i)	• Instructions for assembly, operation and precautions for avoiding excessive laser and collateral radiation.			Provide with the report a sample copy or draft of each user instruction or operation manual which is regularly supplied with the product, or which you cause to be provided with each unit of the reported laser product pursuant to Section 1040.10 (h) (1). *Note:* A sample copy or draft of any new such instruction or operation manual prepared after the submission of this
(h) (1) (ii)	• Description of pulse			

				report must be submitted to the BRH prior to distribution, as a supplement to this report.
(h) (1) (iii)	duration, maximum radiant energy per pulse of assessible laser radiation exceeding Table 7-1A values. • Legible reproductions of all labels and hazard warnings per Section 1040.10 (g) and 1040.11 and location of each label on product.			
(h) (1) (iv)	• Listing of all controls, adjustments and procedures for operation and maintenance.			
(h) (2)	Purchasing and Servicing Information	4.2	All	Purchasing and Servicing Information
(h) (2) (i)	• Catalogs, specification sheets, descriptive brochures and reproduction of warning logotypes.	4.2.1		• Provide with the report a sample copy or draft of each catalog, specification sheet and descriptive brochure containing information pursuant to Section 1040.10 (h) (2) (i).
(h) (2) (ii)	• Instructions for service adjustments and procedures, with warnings and precautions for avoiding excessive laser and collateral radiation; controls and procedures for increasing radiation levels; description	4.2.2		• Provide with report a sample copy or draft of each service instruction or service manual for the reported laser product which you provide or cause to be provided to servicing dealers and distributors and to others upon request containing information pursuant to Section 1040.10 (h) (2) (ii). Note: A sample copy or draft of any new such catalog, specification sheet, descriptive brochure, service instruction, or manual prepared after the submission of the report must be submitted to the BRH prior to distribution, as a supplement to this report.

Table 10-1. (Continued)

Requirements Per Laser Products Performance Standards (Part 1040)		Part & Para. No.	Laser Class	BRH Report Requirements**
Section 1040.10 Para. No.	General Description			Description
	of location of displaceable portions of protective housing: reproduction of labels and hazard warnings.			
—	—	5	—	DESCRIPTION OF THE LASER PRODUCT
—	—	5	—	General: This part is for discussion and reference purposes, and should describe important physical features of the product and identifying names and symbols for these features. • Some of the performance features may be requested in later parts of the report. • Descriptive detail concerning laser and collateral radiation is covered in Part 6.
—	—	5.1	All	Description of Exterior
—	—	5.1.1		• Provide description of exterior of product, which may include photographs or diagrams of all views of product, with parts and component identification and with dimension reference scale.
(b) (12)	Human Access • Means access at a particular point to laser or collateral radiation by any part of the human body, by	5.1.2		• Provide an identifying name (or symbol) and listing of each laser and collateral radiation 'field" or "beam" accessible per definition of human access [Section 1040.10 (6) (12)] from exterior of product.

unobstructed length of 100 cm, or by any line having an unobstructed length of 10 cm, when laser or collateral radiation is incident at that point.

5.1.2.1		—Diagram optical paths of above laser radiation and indicate each aperture for laser radiation.
5.1.2.2		—Indicate on above diagram the location(s) at which the laser radiation parameters requested in Part 6.1 are measured.
5.1.2.3		—Declare which of above radiations must be accessible for purposes of operation, maintenance, service or any combination.
5.1.3		• Provide identifying name (or symbol) and listing of each label affixed to exterior of product, and specify location of each label (can be done in Part 5.1.1).
5.1.4		• Provide identifying name (or symbol) and listing of each control, viewing optic, viewport and display screen that is accessible from the exterior of product.
5.1.4.1		—Specify location of each (can be done in Part 5.1.1).
5.1.4.2		—State when each is intended to be used, for example, during operation, maintenance, service or any combination.
5.1.4.3		—State function of each control.
5.2		Description of Access to Interior
5.2.1	All	• Provide description of each of the means to provide access to interior of product during operation, maintenance, or service (e.g., curtains, doors, ports, service panels, etc.). Can include photographs or diagrams (with dimensional reference scale) of interior as viewed when such means are opened, displaced or removed.
5.2.2		• For each means of providing access to interior, state if it is intended for use during operation, maintenance, service or any combination.

Table 10-1. (Continued)

| Requirements Per Laser Products Performance Standards (Part 1040) | | Part & Para. No. | Laser Class | BRH Report Requirements** |
Section 1040.10 Para. No.	General Description			Description
(b) (12)	Human Access (see definition above)	5.2.3		• Provide identifying name (or symbol) and listing of all laser and collateral radiation which becomes accessible under the definition of human access, by virtue of the means incorporated in the model to provide access to the interior of the product. New designations for previously identified radiation should be introduced only if the radiation has been changed in any characteristic other than direction.
—		5.2.3.1		–Diagram optical paths of above radiation (can be done in Part 5.2.1).
—		5.2.3.2		–Indicate on above diagram the location(s) at which the laser radiation parameters requested in Part 6.1 are measured.
—		5.2.3.3		–State which of above radiation must be accessible for purposes of operation, maintenance, service or any combination.
—		5.2.4		• Provide identifying name (or symbol) and listing of each label affixed to interior of product which becomes visible using means incorporated to provide access to interior.
—		5.2.4.1		–Specify location of each label (can be done in Part 5.2.1).
—		5.2.5		• Provide identifying name (or symbol) and listing of each control, viewing optic, viewport and display screen that becomes operable, usable, functional, etc., by use of the means incorporated to provide access to interior.
—		5.2.5.1		–Specify location of each control, viewing optic, viewport, and display screen (can be done in Part 5.2.1).
—		5.2.5.2		–State when each is intended to be used; e.g., during operation, maintenance, service or any combination.

			Description
—	5.3.1		• Provide description of inner workings of product so reviewer can understand how product operates and achieves its intended function. Also include brief description of any uncertified laser or laser system incorporated into product as to type, manufacturer, model number, power output, etc. Can include photos or diagrams of layout of various parts and components with identification of each, and dimensional reference scale (not needed if provided in Part 5.1 or 5.2).
(b)(12)	5.3.2	Human Access (see definition above)	• Provide identifying name (or symbol) and listing of all laser and collateral radiation which is accessible during operation, maintenance or service, under the definition of human access, in the interior of the product, if such radiation is different from that reported in Part 5.2.3. The identification of the accessible laser radiation should be compatible with that of Part 5.1 and 5.2, and new designations for previously identified radiation should not be introduced.
—	5.3.2.1		—Diagram the optical paths of all laser radiation identified above (can be done in Part 5.2.1).
—	5.3.2.2		—Indicate on above diagram location(s) at which the laser radiation parameters requested in Part 6.1 are measured.
—	5.3.2.2		—State which radiation must be accessible for purposes of operation, maintenance, service or any combination.
(g)	5.3.3	Labeling requirements	• Provide identifying name (or symbol) and listing of each label affixed to interior of product pursuant to Section 1040.10 (g), if not previously identified in Part 5.2.4, and specify location of each label (can do in Part 5.3.1).
—	5.3.4		• Provide identifying name (or symbol) and listing of each control, viewing optic, viewpoint, and display screen in interior of product, if not identified in Part 5.2.5.
—	5.3.4.1		—Specify location of each (can be done in Part 5.3.1).
—	5.3.4.2		—State when each is intended to be used, e.g., during operation, maintenance, service or any combination.
—	5.3.4.3		—State function of each control.

Table 10-1. *(Continued)*

Requirements Per Laser Products Performance Standards (Part 1040)			BRH Report Requirements**
Section 1040.10 Para. No.	General Description	Part & Para. No.	Laser Class	Description

Section 1040.10 Para. No.	General Description	Part & Para. No.	Laser Class	Description
—	—	5.4	—	Nonremovable Laser System. If laser product uses a nonremovable laser system, report changes needed to make it capable of producing radiation when removed. Minor changes (e.g., not requiring any rewiring) are not considered modifications of the laser system. • Thus, laser systems requiring such minor changes to make them capable of producing laser radiation when removed from the product are considered REMOVABLE laser systems and must be certified and reported separately pursuant to Section 1040.10 (c) (2).
(c) (2)	Removable Laser System (see definition given in Part 2.2 above)			
—	—	5.5	All	Descriptions and Diagrams of Circuitry. Provide block diagrams of major electronic and electrical circuitry, including functional descriptions of each block of circuitry. It is necessary to report the detailed circuits within each block when the circuitry PERTAINS TO RADIATION SAFETY, including operational or performance features such as safety interlocks, scanning safeguards, emission duration accumulators, etc.
—	—	5.6	All	Description of Radiation Shielding. Identify, describe and indicate on diagrams requested in Parts 5.1, 5.2 and 5.3 all X-radiation shielding incorporated into products.
—	—	5.7	All	Description of Operation. Describe operation of product and how it achieves intended and known functions (uses) stated in Part 2.6. Describe operation and interrelationship of operation of various parts, components and circuits.

			Description	Required
—	5.8		For simple general purpose laser of well-understood and recognized function and operation, generic name (e.g., "CW He-Ne general purpose laser") will suffice. **Description of Operating Environment** Describe various environments product may operate in order to achieve the intended and known functions (uses) described in Part 2.6. Description should include not only the physical environment of the product (e.g., in extreme conditions of temperature, humidity, shock or vibration), but also the human population likely to be in vicinity of product.	All
(d)	6	Accessible Emission Limits	**REPORTING LEVELS OF ACCESSIBLE RADIATION AND CLASSIFICATION OF THE LASER PRODUCT**	—
	6.1		**Description of Accessible Laser Radiation** Describe accessible laser radiation associated with product and identified in Parts 5.1.2, 5.2.3 and 5.3.2. Sufficient details required to enable the BRH to: • verify the classification of the laser product and • compliance of the product with the applicable requirements of the standard. For laser radiation of more than one wavelength or beam, report each as separately identified laser radiation. Minimum required information follows.	All
(d) (1)	6.1.1	Beam of Single Wavelength	State if wavelength of any of the laser radiation is tunable or adjustable. • If no, state all of the wavelengths in the radiation. • If yes, identify each discrete wavelength (or range) of emission using same identifications as in Parts 5.1.2, 5.2.3 and 5.3.2, giving means by which selection or tuning is achieved.	
(d) (2)		Beam of Multiple Wavelengths In Same Range		
(d) (3)		Beam of Multiple Wavelengths In Different Ranges		
(b) (33)	6.1.2	Scanned Laser Radiation • Laser radiation having time varying direction origin or pattern of propagation with	State if any of the laser radiation is scanned. • If yes: –Identify the scanned laser radiation(s). –Describe the scanning pattern(s). –Give both angular and repetition scanning rates and amplitudes.	

Table 10-1. (*Continued*)

Requirements Per Laser Products Performance Standards (Part 1040)		BRH Report Requirements**		
Section 1040.10 Para. No.	General Description	Part & Para. No.	Laser Class	Description
	respect to stationary frame of reference.			
(f) (9)	Scanning Safeguard			
(c) (1)	Classification of Laser Products–All Laser Products	6.1.3		State classification of product.
(b) (5)	Class I Laser Product • Tables 7-1A, 7-2A, 7-2B			• Class I
(b) (6)	Class II Laser Product • Table 7-1B, 7-2A, 7-2B			• Class II
(b) (7)	Class III Laser Product • Table 7-1C, 7-2A, 7-2B			• Class IIIa or Class IIIb
(b) (8)	Class IV Laser Product • Table 7-1C, 7-2A, 7-2B			• Class IV
—	—	6.1.4		Describe laser radiation characteristics, e.g., • Continuous (CW) • Pulsed • Q-switched • Mode locked • Externally chopped or modulated • Other

—	6.1.4.1	If laser radiation is other than CW, provide the following information:
		(a) Pulse width and criterion for determination (e.g., pulse may be 0.001 second wide at the half-power points).
		(b) Give pulse rise interval and criterion for determination (e.g., pulse may rise in 0.0001 second from 5% power level to the 95% power level).
		(c) Give pulse decay interval and criterion for the interval determination.
		(d) Give maximum pulse repetion rate (PRF)
		(e) Give average energy per pulse and the pulse-to-pulse energy variation from the average.
		(f) Give PRF at which the individual pulse energy is maximum.
		(g) Give PRF at which average radiant power is maximum.
—	6.1.4.2	Describe how above-stated characteristics are achieved.
—	6.1.4.3	Provide diagram or photo of waveform as a linear function of time, including at least the following waveform characteristics:
		• Minimum time interval between pulses
		• Pulse duration
		• Pulse rise and fall time
	6.1.5	State if laser radiation pattern is of such dimension that a significant ($\approx$ 10%) portion of the radiation cannot be collected within the aperture stop specified in Section 1040.10 (e) for measurements of laser radiation.
		If yes:
		–Identify the laser radiation.
		–Describe and diagram pattern.
(e)		Tests for Determination of Compliance
(e) (3)		• Measurement Parameters
(e) (3) (i)		–80 mm diameter circular aperture stop for radiant power (W) or radiant energy (J), except for scanned laser radiation.
(e) (3) (ii)		–7 mm diameter circular aperture stop for

Table 10-1. *(Continued)*

Requirements Per Laser Products Performance Standards (Part 1040)		Part & Para. No.	Laser Class	BRH Report Requirements**
Section 1040.10 Para. No.	General Description			Description
(e) (3) (iii)	irradiance ($W \cdot cm^{-2}$) or radiant exposure ($J \cdot cm^{-2}$) −7 mm diameter circular aperture stop with appropriate solid angle for radiance ($W \cdot cm^{-2} \cdot sr^{-1}$) or integrated radiance ($J \cdot cm^{-2} \cdot sr^{-1}$).			State dominant transverse mode (TEM____). If multi-transverse, give highest order mode anticipated (TEM____). Report and substantiate the class level of each laser radiation identified in Parts 5.1.2, 5.2.3 and 5.3.2, and measured under the conditions described in Section 1040.10 (e).
(e) (4)	• Measurement Parameters for Scanned Laser Radiation −Radiation detectable within a stationary circular aperture stop of 7 mm diameter.	6.1.5.1 6.1.5.2		• BRH suggested (but not mandatory) approach: A thirteen (13)-column table can be filled in containing the following information:
(e)	Tests for Determination of Compliance	6.1.6		

—Column 1　Enter assigned laser RADIATION IDENTIFICA-TION.

—Column 2　Enter CLASS LEVEL of each (composite) laser radiation.

—Column 3　Enter WAVELENGTH(s) of laser radiation in nanometers.

—Column 4　Enter EMISSION DURATION RANGES, in seconds, from:
　　　　　　*Second column of Accessible Emission Limit Tables 7-1A, B, or C for particular class and wavelength(s).

—Column 5　Enter ACCESSIBLE EMISSION LIMIT associated with each of above emission duration ranges from:
　　　　　　*Third column of Accessible Emission Limit Tables 7-1A, B, or C and
　　　　　　*Wavelength-Dependent Correction Factors k_1 and k_2 of Tables 7-2A or B.

—Column 6　Enter largest measured ACCESSIBLE EMISSION LEVEL of radiation for each emission duration of Column 4 above.

—Column 7　Enter MAXIMUM RATIO $\dfrac{E(\lambda, t)}{E(N, \lambda, t)} = \dfrac{\text{Column 6}}{\text{Column 5}}$

—Column 8　Enter specific EMISSION DURATION in seconds, at which the ratio in Column 7 is a maximum.

—Column 9　Enter APERTURE STOP DIAMETER used for measurement of the value reported in Column 6. If other than 7 mm or 80 mm, attach evidence of the equivalency of result.

—Column 10　Enter the identification, as used in Part 10.1, of the RADIATION MEASURING INSTRUMENTS used to measure the radiation level entered in Column 6.

—Column 11　Enter magnitude of ESTIMATED ERROR in mea-sured radiation level reported in Column 6. Note

Table 10-1. (Continued)

Section 1040.10 Para. No.	General Description	Part & Para. No.	Laser Class	BRH Report Requirements** — Description
				that the accessible emission level plus the magnitude of the estimated error must be considered for determination of Class level.
				—Column 12 State if radiation is necessary for the performance of the function(s) of the product (Yes or No).
				—Column 13 State if access to radiation is necessary during operation of product, or if access to radiation is necessary only during service of product.
				• Alternate methods such as graphical presentation may be used to report the information, provided all of the requested information is included.
				Notes: (1) Each of the accessible laser radiations may consist of more than one wavelength, and may be considered as a composite radiation.
				• It is composite radiation which is evaluated using the requirements of Section 1040.10 (d) and with which a class level can be assigned.
				(2) The purpose of testing a particular wavelength component of laser radiation against accessible emission limits is to determine within an emission duration range the highest level of that component.
(d)	Accessible Emission Limits	6.2	—	
—				Reporting Collateral Radiation
				State if product produces any accessible optical collateral radiation.
				• If yes, proceed to Part 6.2.1.

(d)	6.2.1	—

Table 7-1A Accessible limits for Collateral Radiation from Laser Products.
• Same as Class I laser radiation using Table 7-1A

(h) Informational Requirements

(h)(1) • User Information
(h)(2) • Purchasing and Servicing Information

Optical Collateral Radiation (λ = 250 to 13,000 nm)

Describe optical collateral radiation associated with the product and identified in Part 5.

Detail should be such that BRH reviewer can ascertain:

• If accessible radiation level is either below or above the accessible emission limits of Table 7-1A of Section 1040.10 (d).

• Whether the operation, maintenance and service instructions are adequate to meet the requirements of Section 1040.10 (h).

For each identified collateral radiation, at least the following information is required, and may be presented in a nine-column tabular summary:

• Column 1   Enter assigned collateral RADIATION IDENTIFICATION.

• Column 2   Enter the origin (source) of the field (LOCATION OF ORIGIN OF RADIATION) using the identification for locations in Part 5.

• Column 3   Identify the physical nature of the source of the radiation (TYPE OF RADIATION); e.g., high temperature thermal-plankian radiator, pump source emission, fluorescence, plasma glow, etc.

• Column 4   Enter WAVELENGTH RANGE(s) emitted, in nanometers.

• Column 5   Enter EMISSION DURATION RANGES, in seconds, from:

    —Second column of Table 7-1A (same as Table 7-1A for Class I Accessible Emission Limits), for particular wavelength(s) or range(s)

• Column 6   Enter HIGHEST ACCESSIBLE EMISSION LEVEL for each emission duration range.

Table 10-1. (*Continued*)

Requirements Per Laser Products Performance Standards (Part 1040)				BRH Report Requirements**
Section 1040.10 Para. No.	General Description	Part & Para. No.	Laser Class	Description
				–If source is a continuum or is extended, a spectral output curve may be attached. –If levels are above accessible emission limits of Table 7-1A, it is only necessary to indicate which ones are above those limits. • Column 7 Enter ESTIMATED ERROR in the measured radiation level reported in Column 6. • Column 8 Enter MEASURING INSTRUMENTATION used to determine radiation value of Column 6. • Column 9 Enter the location at which the measurement was made using the position identifications prepared in Part 5 (MEASUREMENT LOCATION IDENTIFICATION).
—	—	6.2.2	—	Collateral X-Radiation State if product emits a measurable amount of X-radiation. (In general, products with operating voltages less than 5 KV will not emit measurable amounts of X-radiation) If yes, respond to following:
		6.2.2.1		• Provide sources or possible sources of the X-radiation.
		6.2.2.2		• Provide: –Maximum operating voltages (not starting voltages for gas discharges) of these sources. –Purposes of each of these voltages (e.g., energy storage, triggering, particle acceleration, maintenance of gas discharge, etc.).

6.2.2.3

- Provide the maximum accessible X-ray emission in milliroentgens (mR), averaged over an area of 10 cm² having no dimension greater than 5 cm and parallel to the external surface of the product in one hour of product operation at those operating conditions which maximize X-ray emission.

6.2.2.4

- Describe how product was surveyed to determine location of maximum X-ray emission and identify location.

6.2.2.5

- Describe method and apparatus of the measurement of X-ray exposure; include probable errors.

6.2.2.6

- Provide features of product (e.g., distance of source from external surface, interposed shielding, power regulation) that maintain level of collateral X-radiation below the accessible emission limits of Table 7-1A Section 1040.10 (d).

(d) Table 7-1A Accessible Emission Limits For Collateral Radiation From Laser Products 2. Accessible emission limit for X-radiation.

6.3 —

State if the accessible levels of laser and collateral radiation have been determined under the conditions which maximize both the level of accessible emissions and the detection of such emissions.

(f) PERFORMANCE REQUIREMENTS —

COMPLIANCE WITH THE PERFORMANCE REQUIREMENTS

7.1 Protective Housing (laser radiation)

(f) (1) (i) (ii) (iii) II, III, IV

Protective Housing (laser radiation)
For each accessible laser radiation during operation (per Part 5), explain why level is within the lowest class that can be used for the performance of the function(s) of the product.

7.2 Protective Housing (collateral radiation)

(f) (1) (iv) All

Protective Housing (collateral radiation)
Justify that any accessible collateral radiation during operation (per Part 5) in excess of limits of Table 7-1A must be accessible in order to perform the intended function(s) of the product.

7.3 Safety Interlocks

(f) (2) (i) All

Safety Interlocks

7.3.1
- Provide detailed electrical and mechanical diagram of each safety interlock (per Part 5) used for radiation safety.

7.3.2
- For each safety interlock state if actuation is intended during operation, maintenance, service or any combination.

Table 10-1. (*Continued*)

Requirements Per Laser Products Performance Standards (Part 1040)		BRH Report Requirements**		
Section 1040.10 Para. No.	General Description	Part & Para. No.	Laser Class	Description
		7.3.3		• State highest level of laser and collateral radiation to which access is prevented for each safety interlock not designed to allow defeat.
		7.3.4		• Describe how each safety interlock prevents access to laser or collateral radiation in excess of the accessible emission limits per Section 1040.10 (f) (1) upon removal or displacement of interlocked portion of protective housing. Includes interlocks designed to be defeated.
(f) (2) (ii) (iii)	Safety Interlocks	7.4 7.4.1 7.4.2	All	Defeatable Safety Interlocks • State which safety interlocks are designed to allow defeat. • State if defeat is intended during operation, maintenance, service or any combination.
		7.4.3		• Describe why replacement of a removed or displaced portion of the protective housing is not possible while the required safety interlocks are defeated.
		7.4.4		• Describe the mechanical, electrical and operational characteristics of the means used to provide a visual or an aural indication of defeat.
		7.4.5		• Confirm that during safety interlock defeat the indication of such defeat is visible or audible whenever the laser product is energized, and that the meaning of the indication is readily apparent.
(f) (2) (i) (b)	Safety Interlocks	7.5 7.5.1	All	Safety Interlock Failure • Describe how each safety interlock precludes removal or displacement of the appropriate interlocked portion of the protective

housing upon failure of the safety interlock to prevent access to laser or collateral radiation.

7.5.2		• Describe the possible modes of failure of each safety interlock and the resultant effect upon the radiation safety of the laser product.
7.5.3		• State number of operational cycles before failure.
7.5.4		• For each safety interlock state number of actuations expected to occur on the average during each day of operation of the laser product.
(f) (3)	Remote Control Connector	III, IV
7.6	Remote Control Connector	
7.6.1		• Provide detailed description of electrical and mechanical construction.
7.6.2		• Describe functioning (operation) of remote control connector and how it prevents access to laser and collateral radiation from the laser product in excess of the accessible emission limits of Class I and Table 7-1A of Section 1040.10 (d) when terminals of connector are not electrically joined.
7.6.3		• State the open-circuit electrical potential difference on the terminals of the remote control connector ($\leqq$ 130 volt RMS required).
7.6.4		• State whether the function fulfilled by the remote control connector is available at all times during operation, maintenance and service, regardless of any other requirement of Sections 1040.10 and 1040.11.
(f) (4)	Key Control	III, IV
7.7	Key Control	
7.7.1		• Provide description of electrical and mechanical construction of key-actuated master control.
7.7.2		• Describe functioning (operation) of key-actuated master control and how it renders laser inoperable when key is removed.
7.7.3		• State whether the function fulfilled by the key-actuated master control is available at all times during operation, maintenance and service regardless of any requirement of Sections 1040.10 and 1040.11.
(f) (5)	Laser Radiation Emission Indicator	II, III, IV
7.8	Laser Radiation Emission Indicator	
7.8.1		• State whether the laser and laser energy source are housed separately.

Table 10-1. (Continued)

| Requirements Per Laser Products Performance Standards (Part 1040) | | | BRH Report Requirements** |
Section 1040.10 Para. No.	General Description	Part & Para. No.	Description
		7.8.2	• If yes, —State if they can be operated at a separation distance of > 2 meters and if any special equipment is needed to operate at that separation distance. —Describe the emission indicators on both the laser and laser energy source in Parts 7.8.3 through 7.8.11.
		7.8.3	• Describe in detail the mechanical and electrical characteristics of all installed emission indicators.
		7.8.4	• Describe the functioning (operation) of the indicators.
		7.8.5	• State whether emission indicator provides visible or audible signal during emission of accessible laser radiation that is in excess of the accessible emission limits of Class I.
		7.8.6	• Describe the compliance of your product with Section 1040.10 (f) (5) (iv) (e.g., protective eyewear must allow viewing of any visible emission indicator).
		7.8.7	• State whether the meaning of the signal from the emission indicator is readily understandable by individuals and explain why this is so.
		7.8.8	• Confirm here and show in Part 5 that the emission indicators are located so that viewing does not require human exposure to laser or collateral radiation in excess of the accessible emission limits of Class I and Table 7-1A.
		7.8.9	• State whether the function fulfilled by the radiation emission indicator is available at all times during operation, maintenance

and service, regardless of any other requirements of Sections 1040.10 and 1040.11.

	7.8.10		• State the length of time each emission indicator is actuated prior to the emission of accessible laser radiation in excess of Class I.
	7.8.11		• For Class III or IV laser systems, justify that the lengths of time stated above are each sufficiently long to allow appropriate action to avoid exposure to laser radiation.
			Note: Identification and location of emission indicators should be given in Part 5.
(f) (5) (iv)	7.9	II, III, IV	Protective Eyewear
			State whether protective eyewear is supplied with or recommended for use with the laser system.
			If yes:
	7.9.1		• Provide spectral transmission at wavelength(s) of laser and collateral radiation emission(s), and also durability specifications of protective eyewear.
	7.9.2		• Confirm that protective eyewear transmits sufficient luminous flux at those visible light wavelengths necessary to allow visibility of emission indicators.
(f) (6)	7.10	II, III, IV	Beam Attenuator
	7.10.1		• Describe in detail the mechanical and electrical characteristics of each permanently attached means (e.g., beam attenuator) other than the laser energy source switch(es), electrical supply main connectors or the key-actuated master control, capable of preventing human access to all laser and collateral radiation in excess of the limits of Class I and Table 7-1A.
	7.10.2		• Discuss the permanency of attachment of each beam attenuator.
	7.10.3		• Discuss and confirm with data how each beam attenuator, when actuated, prevents human access to all laser and collateral radiation in excess of the accessible emission limits of Class I and Table 7-1A.
	7.10.4		• Confirm that the function fulfilled by the beam attenuator is available at all times during operation, maintenance and service, regardless of any other requirement of Sections 1040.10 and 1040.11.

Table 10-1. (Continued)

Requirements Per Laser Products Performance Standards (Part 1040)		BRH Report Requirements**		
Section 1040.10 Para. No.	General Description	Part & Para. No.	Laser Class	Description
(f) (7)	Location of Controls	7.11	II, III, IV	Location of Controls Discuss how the location of each of the operation and adjustment controls of the laser product is such that human exposure to laser or collateral radiation in excess of the accessible emission limits of Class I and Table 7-1A is unnecessary for operation or adjustment of such controls. *Note:* Identification and location of controls relative to accessible laser and collateral radiation should be given in Part 5.
(f) (8)	Viewing Optics	7.12 7.12.1	All	Viewing Optics • State whether any (all) laser and collateral radiation that is accessible by virtue of viewing optics, viewports, and display screens incorporated into the laser product is at all times not greater than the accessible emission limits of Class I and Table 7-1A.
		7.12.2		• Describe in detail with diagrams or photographs and radiation transmission or reflection spectra, each shutter or variable attenuator incorporated into viewing optics, viewport or display screen.
		7.12.3		• Describe the mechanical, electrical, and functional characteristics of the means incorporated into the laser product by which access is prevented to laser or collateral radiation in excess of the accessible emission limits of Class I and Table 7-1A.
		7.12.4		• Describe how opening the shutter or varying the attenuation is precluded when human access to transmitted laser or collateral radiation in excess of the accessible emission limits of Class I

and Table 7-1A is possible upon failure of such means as required in Section 1040.10 (f) (8) (i) (e.g., Part 7.12.3 above).

Note: Location and identification of laser and collateral radiation made accessible by virtue of viewing optics, viewports and display screens should be given in Part 5, and their highest level set forth in Part 6.

Ref.	Item	Section	Class	Description
(f) (9)	Scanning Safeguard	7.13	All	Scanning Safeguard (for lasers with scanned laser radiation) Describe the mechanical, electrical and functional characteristics of means by which human access is not permitted to laser radiation in excess of the accessible emission limits (that are applicable to the scanned radiation) as a result of scan failure or other failure causing a change in either scan velocity or amplitude of the laser radiation.
1040.11 (a)	Medical Laser Product	7.14 / 7.14.1	III, IV	Medical Laser Product • Describe in detail the means incorporated into the product to measure the level of laser radiation intended for irradiation of the human body, including circuit diagrams and optical system diagrams for the measurement system.
		7.14.2		• Specify the estimated uncertainty in the measurement system and the method by which it was derived (measurement error must be no more than ± 20%
		7.14.3		• Provide a copy of the instructions specifying the procedures and schedule for calibration of the measurement system required to be incorporated in Class III or IV medical laser products.
(e)	Tests For Determination of Compliance	8	All	QUALITY CONTROL TESTS AND TESTING PROCEDURES FOR COMPLIANCE Provide a summary of quality control tests and procedures: • Used in prototype, preproduction and initial model development • For components • In production • In retesting of rejected units
(e) (1)	• Tests for Certification			
(e) (2)	• Test Conditions			
(e) (3)	• Measurement Parameters			
(e) (4)	• Measurement Parameters for Scanned Laser Radiation	8.1	—	Prototype, Preproduction and Initial Model Development Tests Describe any applicable tests (except life and endurance tests of

Table 10-1. *(Continued)*

Requirements Per Laser Products Performance Standards (Part 1040)				BRH Report Requirements**
Section 1040.10 Para. No.	General Description	Part & Para. No.	Laser Class	Description
—	—			Part 9) used during prototype, preproduction or initial model development, • only when such tests are used to assure compliance of subsequent production and are relevant to the final production model. Each test description should include: • Statement of those features tested • Parameters measured • Test procedures • Test equipment and setups including diagrams, photos and a clear identification of each test instrument. Where applicable such tests shall include:
—	—	8.1.1		• Tests of laser radiation emission even if it is not expected to be accessible in the final product.
—	—	8.1.2		• Any tests to evaluate performance of interlocks, remote control connectors, key controls, laser radiation emission indicators, scanning safeguards and any other radiation safety features whose performance may be measured directly or indirectly when tested during production.
—	—	8.1.3		• Any tests of beam attenuators, viewing optics, filters, reflectors, lenses or other optical components or surfaces that may affect the accessible radiation level from the final product.
—	—	8.1.4		• Any tests to measure the emission of collateral radiation including the levels and spectral quality of emissions.
—	—	8.1.5		• Any tests relative to effects of operating or environmental conditions such as variations with live voltage, sensitivity to operating

8.1.6	temperatures, variations with cooling rates, effects of use and abuse, etc. • Any other tests conducted or to be conducted that are believed to confirm or document the radiation safety of the product and its compliance with the portions of 21 CFR Chapter I, Subchapter J, applicable to laser products.
8.2	Component Tests Describe quality control and testing procedures for components affecting radiation safety, conducted prior to their installation in the laser product, that are considered to be a vital and necessary part of the testing program to assure compliance with the Federal performance standard. Procedures shall include at least: • Incoming inspection • And/or subassembly testing of such items as power control and monitor controls, emission indicators, interlock switches, scanning mechanisms, etc. Where applicable the descriptions shall include:
8.2.1	• Vendor qualification requirements.
8.2.2	• Incoming inspection procedures, accept/reject criteria and lot and sample size if not 100% tested. If 100% tested, so state.
8.2.3	• Subassembly test procedures, parameters and limits, and lot and sample size if not 100% tested. If 100% tested, so state.
8.2.4	• Describe corrective action procedures following unit or lot rejection.
	Type A information (Testing Procedures – conducted on 100% basis) For those tests listed under Parts 8.3 and 8.4 the following information should be given for each test:
	• A1 Statement of which features are tested and which parameters are measured or monitored.
	• A2 Description of procedures followed.
	• A3 Description of test equipment and setups with diagrams, photos and a clear identification of each test instrument.

Table 10-1. *(Continued)*

Requirements Per Laser Products Performance Standards (Part 1040)			BRH Report Requirements**
Section 1040.10 Para. No.	General Description	Part & Para. No.	Laser Class

- A4 Physical and electrical conditions under which tests are made (e.g., live voltage, operating temperature, cooling rate, etc.).

- A5 Adjustments, if any, made during tests and specific procedures for making adjustments.

- A6 Description of accessible laser and collateral radiation measurement procedures if applicable, including detector scanning pattern, surfaces and areas surveyed for radiation emission and position of detector from prominent surface and means of maintaining this position.

- A7 Test acceptance and rejection criteria (e.g., emission limits, tolerances on interlocks, emission indicators, beam attenuators and other safety performance features).

- A8 Reference (by document title and number, page and paragraph or test station) to any applicable portions of internally written quality control documents, testing procedures and flow charts. Attach copies of such documents to this report. If there are no such documents, so state.

- A9 State when during the production process the test is conducted, and whether the test is a final test or whether it is repeated at a subsequent stage of production and testing (other than by sample or audit).

- A10 Describe the records that are made and retained regarding these tests.

Type B Information (Sampling Procedures)

For those tests listed under Part 8.3, which are not conducted on a 100% basis, the following information should be given for each test.

• B1 Reference to any published acceptance sampling plans.
• B2 Description of sampling schemes used including sample size, lot size from which sample is taken, and rejection/acceptance criteria for sampled units and sampled lots.
• B3 Proportion of total production inspected.
• B4 Rationale for determining sample size.
• B5 Method of sample selection to assure randomness.
• B6 Correction action which would be taken subsequent to unit rejection including but not limited to temporary changes in sampling plan, inspection of the rejected lot and reinspection of lots produced subsequent to and prior to rejected lots.

8.3

Production Tests

Describe tests and testing procedures (exclusive of those of Parts 8.1 and 8.2) conducted during and after completing the manufacturing process (regardless of whether or not the test is repeated during final test).

This includes installation of the laser product, including check-out for proper operation and function.

Description of the tests, if performed, should include but not be limited to the following:

8.3.1

• Tests of laser radiation emission that would be accessible during operation, maintenance or service in the final product.
• Tests and adjustments of the following:

8.3.2
8.3.2.1   –Safety interlocks.
8.3.2.2   –Visual or aural indicators of interlock defeat.
8.3.2.3   –Remote control connectors.
8.3.2.4   –Key controls.
8.3.2.5   –Laser radiation emission indicators.
8.3.2.6   –Beam attenuators.
8.3.2.7   –Viewing optics,
8.3.2.8   –Scanning safeguards.

Table 10-1. *(Continued)*

Requirements Per Laser Products Performance Standards (Part 1040)			BRH Report Requirements**
Section 1040.10 Para. No.	General Description	Laser Class	Description
1040.11 (a) 1040.10 1040.11	Medical Laser Products Laser Products Specific Purpose Laser Products		—Radiometric instruments pursuant to Section 1040.11 (a). —Fail-safe devices required by or made necessary in order to comply with Sections 1040.10 and 1040.11.
8.3.3			• Tests and adjustments of protective housing panels, doors, latches, and so forth.
8.3.4			• Tests and adjustments of filters, reflectors, lenses or other optical components that may affect the level of accessible radiation.
8.3.5			• Tests of emission of collateral radiation including the levels and spectral quality of emissions.
8.3.6			• Tests and adjustments concerning operating or environmental conditions such as operating voltages, operating temperatures, cooling rates, etc.
8.3.7			• Any other tests or adjustments are made and believed to confirm or document the radiation safety of the product or its compliance with Sections 1040.10 and 1040.11.
8.4 1040.10 1040.11	Laser Products Specific Purpose Laser Products	—	Retesting of Rejected Units Describe the procedure and location for retesting of all units rejected in tests described in Part 8.3.
8.4.1			• Where retesting is accomplished by rerouting the rejected product back into a production line, indicate the points (relative to testing stations) at which the product reenters the line. If anything less than 100% retest is performed on the reworked unit, detail the omitted tests.

8.4.2		• Where testing is performed off the production line provide all applicable information of "Type A Information" and "Type B Information."
8.5		Results of Tests Summarize the results of all tests reported in Part 8. This should include at least the following information for each type of test procedure performed:
8.5.1		• Identification of type of tests for which results are presented including reference to applicable portions of Parts 8.1 through 8.3 as appropriate.
8.5.2		• Time period represented by results presented.
8.5.3		• Number of units tested.
8.5.4		• Percent of proportion of total units manufactured or received (in the case of components) that were tested.
8.5.5		• Where sufficient data are available, the mean, range and standard deviation of each type of measurement. If these values are unavailable, other representative statistics or expressions of results may be reported.
8.5.6		• Where test results are recorded as numeric values, provide a frequency distribution in the form of a graph, or chart, or other suitable means to convey the same information.
8.5.7		• If tests are recorded on a go/no-go basis, give the number or units which failed the test and percentage which failed the test. If lot sampling was employed, also provide number and percentage of failed lots.
9	All	LIFE AND ENDURANCE TESTING
9.1	—	Test Procedures Describe any life and endurance testing procedures used to date, and that will be used in future testing of the laser product, to determine the ability of the product to comply with the performance standard throughout its useful life.
(e)		Tests For Determination of Compliance
(e) (1)		Tests for Certification • Because compliance with the standard is required for the useful life of a product,

Table 10-1. (*Continued*)

Requirements Per Laser Products Performance Standards (Part 1040)			BRH Report Requirements**
Section 1040.10 Para. No.	General Description	Part & Para. No.	Laser Class	Description

Section 1040.10 Para. No.	General Description	Part & Para. No.	Laser Class	Description
	such tests shall also account for increases in emission and degradation in radiation safety with age.	9.1.1		In the description: • Identify all parts, components and accessories —whose deterioration with age or use could cause increased level of radiation or —otherwise degrade the radiation safety of the product and which could cause the product to be in noncompliance with the standard. • Give method and estimate size of any increase in emissions or degree to which radiation safety of product is compromised by deterioration of part, component or accessory. • If operating life or endurance of such parts, components or accessories is included in vendor specifications, state which are purchased and corresponding specifications. • Endurance and reliability test may include at least:
		9.1.2		—Radiation output from the laser and variation with age or use.
		9.1.2.1		—Safety interlocks.
		9.1.2.2		—Remote control connectors.
		9.1.2.3		—Key controls.
		9.1.2.4		—Laser radiation emission indicators.
		9.1.2.5		—Beam attenuators.
		9.1.2.6		—Viewing optics.
		9.1.2.7		—Scanning safeguards.
		9.1.2.8		
1040.11 (a)	Medical Laser Products	9.1.2.9		—Radiometric instruments pursuant to Section 1040.11 (a).
		9.1.2.10		—Any other parts, components or accessories that might affect the radiation safety of the reported product.

9.1.2.11 — The effect upon radiation safety of build-up of foreign substances on beam attenuators, scanning safeguards, mirrors, filters, viewing optics and other safety features.

9.1.3 • The following information is required, as applicable, for each test conducted and reported in Part 9.1.2.

9.1.3.1 — Description of procedures followed.

9.1.3.2 — Number of units tested, frequency of such testing and any selection criteria for the units tested.

9.1.3.3 — Parameters monitored, measured or tested.

9.1.3.4 — Physical and electrical conditions under which tests are made (e.g., live voltage, ambient temperature, cooling rates, etc.)

9.1.3.5 — Adjustments, if any, made during tests and how made.

9.1.3.6 — Instruments and test equipment for each test including description of any special test apparatus or devices.

9.1.3.7 — Description of laser or collateral radiation emission measurement procedures including detector scanning pattern, surfaces and area surveyed for radiation emission and distance of detector from laser product surface.

9.1.3.8 — Time and number of operational cycles between tests with a definition of what constitutes an operational cycle.

9.1.3.9 — Test result criteria for acceptance or rejection of the product.

9.1.3.10 — Reference by document title and number, page and paragraph any applicable portions of internally written quality control documents and testing procedures. Attach copies of such documents as an appendix to this report.
If there are no such documents, so state.

9.1.4 • Provide corrective actions that are or will be taken after rejection of a part, component, accessory or unit.

9.2 Results of Life and Endurance Tests
Summarize or individually report results of life and endurance tests conducted on reported model of laser product.
Include at least following information for each type of test procedure performed:

Table 10-1. (*Continued*)

Requirements Per Laser Products Performance Standards (Part 1040)			BRH Report Requirements**
Section 1040.10 Para. No.	General Description	Part & Para. No.	Laser Class	Description
		9.2.1		• Identification of type of test for which results are presented including reference to applicable portions of Part 9.1.
		9.2.2		• Time period represented by results presented.
		9.2.3		• Number of units tested and percent or proportion of total units manufactured that were tested.
		9.2.4		• For each unit tested provide:
		9.2.4.1		—Dates of first and last test.
		9.2.4.2		—Results of each type of test at the beginning of test period, at the end of test period and at all other times during the test period when the test was conducted.
		9.2.5		• Component failures, time of failure or number of operational cycles to failure, means of correction and effects of correction on laser or collateral radiation and any other safety characteristics of the laser product.
	—	9.3	—	Malfunctions in Laser Product Provide information on common causes for any malfunction in laser product that would change its radiation safety performance characteristics or ability to comply with the federal standard. Report the number of malfunctions that have occurred. • Describe any safety features incorporated in product or safety procedures recommended to prevent injury in case of such malfunction.
	—	9.4	—	Repair or Replacement Problems State what repair or replacement problems have occurred (if any) that affect the radiation safety of the laser product.

(e)	Tests for Determination of Compliance	10	All
1040.10	Laser Products	10.1	—
1040.11	Specific Purpose Laser Products		

INSTRUMENTATION AND CALIBRATION

Instrumentation and Testing Setups

Describe all instruments and testing setups used to assure compliance of reported laser product with requirements of Sections 1040.10 and 1040.11.

Include instruments and testing setups for:
- Engineering prototypes
- Preproduction units
- Initial production units

Reference may be made to instrumentation and testing setups described in Parts 8 and 9.

10.1.1
- Identify each instrument or testing setup as to
  - Manufacturer, model number and date of acquisition.
  - Whether it is used in engineering, production or quality control programs.

10.1.2
- Provide diagrams and photos of test setups, including parts identification of each instrument and/or optical component, if such are helpful in describing test setups.

10.1.3
- Specify, where applicable, for each instrument or test setup the:

10.1.3.1 — Measurement amplitude range and estimated error.

10.1.3.2 — Response time.

10.1.3.3 — Effective measurement area or aperture stop size.

10.1.3.4 — Type and characteristic of detector.

10.1.3.5 — Wavelength dependence characteristics (particularly for the detector and the entire instrument or test setup).

10.1.3.6 — Linearity of response (particularly, for instruments used in measuring beam power; the linearity over wide ranges of intensity and duration of incident radiation, such as between short, intense pulses and weak, continuous calibration sources).

10.1.3.7 — Angular dependence of detector (or of the instrument or test setup).

10.1.3.8 — Temperature dependence of detector (or of instrument or test setup).

Table 10-1. (*Continued*)

Requirements Per Laser Products Performance Standards (Part 1040)			BRH Report Requirements**
Section 1040.10 Para. No.	General Description	Laser Class	Description
10.1.3.9			–Solid angle of acceptance having a circular cross section.
10.1.3.10			–Uniformity of response over detector sensitive cross section (or instrument or test setup).
10.1.4			• For each measurement uncertainty reported in Parts 6.1, 6.2 and 6.3 for accessible laser and collateral radiations, identify, here:
			–the measuring instrument system and
			–present the method and particulars by which the uncertainty was determined.
			Demonstrate:
			–how each uncertainty is propagated in the measurement process and
			–the method used in the combining of uncertainties into the overall estimate reported.
10.1.5			• If the reported product is not introduced into commerce (supplied) with its own laser, specify, by brand name and model number, what laser is used during testing.
10.1.6			• If product is not supplied with its own laser energy source (power supply), specify, by brand name and model number, what source is used during testing.
10.1.7			• State for each instrument or testing setup what parameters are measured and relate each parameter to the appropriate specific requirement of the standard.
10.1.8			• Provide description of test or measurement procedure used for

10.1.9	each parameter (if this was not done in Parts 8 or 9) and how any derived parameters are calculated from direct measurements. • Describe by model, use location and serial number, if applicable, any modifications made to any of the above-listed instruments if such instruments are available commercially, and describe in detail any such modifications.
—	
10.2	Checks of Instrumentation Describe what checks of instrumentation are performed.
10.2.1	• State what tests and calibrations of instrumentation are made and how often each is made. Describe any criteria used for determining if recalibration is necessary.
10.2.2	• If some calibration of instruments is done by someone else (calibration lab, another company, and so forth), state who does the calibration of which instruments and how often it is done for each instrument.
10.2.3	• State what apparatus is used to check or calibrate the instrumentation. List all equipment, including manufacturer, model and relevant specifications, used to make the check or calibration.
10.2.4	• Provide schematic or block diagrams of the apparatus and test setups employed in checking and calibrating the instruments.

**Per Reference No. 5 "Guide For Submission of Information on Lasers and Products Containing Lasers Pursuant to 21 CFR 1002.10 and 1002.12."

adapted from Reference 5, and contains an example of CW radiation of multiple wavelengths.

The BRH issued a letter dated July 29, 1976 setting aside portions of its standard for laser products built for combat training or combat operations of the Department of Defense.[8] Laser systems that are exempt include:

- Laser rangefinders and target designators
- Aircraft displays and eye-pointing indicators for gunners or pilots
- Field training lasers (e.g., direct-fire simulators)
- Classified military laser systems

All other lasers used by the Department of Defense for industrial, classroom, construction, alignment, research, instrumentation, and medical purposes, and so on, are not exempted from the BRH code.

Military lasers excluded from BRH standards must prominently display the following label:

CAUTION

This electronic system has been exempted from FDA radiation safety performance standards prescribed in the Code of Federal Regulations, Title 21, Chapter 1, Subchapter J, pursuant to exemption No. 76 EL-01 DOD issued on 26 July 1976. This product should not be used without adequate protective devices or procedures.

Two additional requirements exist for exempted military laser systems:

- The Department of Defense must provide an annual report to BRH of the types and quantity of lasers or laser products obtained or released under this exemption.
- Manufacturers of such exempted lasers are required to report laser accidents to BRH, as is the case for lasers that are not exempted.

User Safety Standards

As noted earlier, OSHA is in the process of developing broad laser user standards, while the Bureau of Labor Standards has in effect a relatively minor laser user standard for construction work. On the basis of relative importance, the presently known OSHA requirements will be first addressed.

Table 10-2, adapted from the *Laser Focus Buyers Guide*,[8] provides a comparison of 21 key laser safe use control measures as presently contained in ANSI's[1] voluntary standard and/or OSHA's proposed standard, or not covered in either standard. The BRH hazard classification scheme (Class I–IV) is also to be used by OSHA.

A summary of laser safe use control measures required in the Bureau of Labor Standard's "Safety and Health Regulation for Construction"[7] is contained in Table 10-3.

STATE LASER SAFETY LEGISLATION

Concern with laser safety at the state level is considerable. A total of 31 states have some form of state laser regulations either passed, pending or empowered by enabling legislation or OSHA-state agreements.[10] This section will examine on a survey basis the variations in status and/or regulations for those 31 states. Then, detailed emphasis will be given to an interpretation of New York State Industrial Code Rule 50, for the safe use of lasers.

Survey of Existing and Proposed State Laser Safety Legislation

Table 10-4, adapted from *Laser Institute of America News*,[10] shows five groupings of the 31 states with some form of laser safety obligation.

The first eleven states of Table 10-4 have some type of regulations or voluntary laser safety program.

Six additional states (12–17 of Table 10-4) have enacted enabling legislation that empowers the state health departments or equivalent agencies to develop some type of laser safety standards. Draft legislation is in preparation or has been prepared for each of these states.

Table 10-2. Laser Use Control Measures Required by ANSI and OSHA Standards.[1]

No.	Safe Use Requirement	Class I		Class II		Class IIIa		Class IIIb		Class IV	
		ANSI	OSHA	ANSI	OSHA	ANSI	OSHA	ANSI	OSHA	ANSI	OSHA
1	Use properly marked eye protection when eye exposure possible.	N/A	N/A		[2]	X	X	X	X	X	X
2	Do not use system if laser protective filters appear damaged.		X		X		X		X		X
3	Only authorized operators.[3]										
4	Viewing optics must attenuate radiation to MPE or Class I levels.		X		X	X	X	X	X	X	X
5	No tracking of nontarget vehicles.	N/A				X	X	X	X	X	X
6	Eye exposure limited to ocular MPE.	N/A	N/A		X	X	X	X	X	X	X
7	Skin exposure limited to skin MPE.	N/A	N/A	N/A	N/A	X		X		X	
8	Training of laser user.					X		X		X	
9	Reduce laser output to Class I levels when beam not in use.				X		X		X		X
10	Use laser in "controlled area."							X		X	
11	Limit to spectators.							X		X	
12	Visible or audible emission indicators.						X		X		X
13	Remove unessential specular surfaces from laser beam path.							X	X	X	X
14	"Caution" warning tags and labels.			X	X	X	X				
15	"Danger" warning tags and labels.							X	X	X	X
16	Restricted beam access and barriers.							X	X	X	X
17	Closed installation when indoors.										X
18	Safety power shutoff switch.								X		X
19	Posting of danger signs in laser area.								X	X	X
20	Remote firing and monitoring advisable.									X	
21	Key switch master interlock.					X		X			

[1] Adapted from Reference 8.

**Table 10-3. Construction Laser Use Control Measures Required
by the Bureau of Labor Standards.[1]**

Item No.	Section No.	Safe Use Requirement
1	1518.54 (j)	Exposure limits: (1) Direct staring: $\leqq 1 \times 10^{-6} \text{W} \cdot \text{cm}^{-2}$ (2) Incidental observing: $\leqq 1 \times 10^{-3} \text{W} \cdot \text{cm}^{-2}$ (3) Diffused reflected light: $\leqq 2\frac{1}{2} \text{W} \cdot \text{cm}^{-2}$
2	1518.54 (g)	Laser beam not to be directed at individuals.
3	1518.54 (k)	When possible, laser unit in operation should be set up above the heads of individuals.
4	1518.54 (c)	Use antilaser eye protection (e.g., safety goggles) when potential exposure to direct or reflected laser light $>5 \times 10^{-3}$ watt exists.
5	1518.54 (a)	Only qualified and trained individuals are to install, adjust and operate laser equipment.[2]
6	1518.54 (b)	Proof of qualification as laser equipment operator to be available and in possession of operator at all times.
7	1518.54 (h)	Laser operation should be prohibited where possible, during rain, snow and when fog or dust is in the air.
8	1518.54 (e)	Use beam shutters or caps, or turn laser off, when laser transmission not required. Turn laser off when left unattended for substantial periods of time (e.g., lunch, shift change, overnight, etc.)
9	1518.54 (f)	Only mechanical or electronic means are to be used as a detector for guiding the internal alignment of the laser.
10	1518.54 (d)	Post areas with standard laser warning placards when laser is used.
11	1518.54 (i)	Label required on laser equipment to indicate maximum output (of laser).
12	1518.102 1518.102 (a) (2) 1518.102 (b) (2) (ii)	Eye Protection Specifics (a) Eye protection equipment shall meet requirements specified in ANSI's Z 87.1-1968.[3] (b) Protective goggles require label with following information: • Laser wavelength for which use is intended. • Optical density of these wavelengths. • Visible light transmission.

[1]Compiled from Reference 7, "Safety and Health Regulations For Construction."
[2]Qualification and training requirements not specified, as for example in New York State Industrial Code Rule 50 "Lasers."
[3]See Reference 9.

Table 10-4. States With Some Form of Current Laser Safety Obligation.[1]

Category Description	No.	State	Agency	Title	Date
States with current regulations—including registration	1	Alaska	Dept. Envir. Conserv.	Title 18, Art 7 and 8	10/71
	2	Georgia	Dept. Health	Ch 270-5-27	9/1/71
	3	Illinois	Dept. Pub. Health	Registration Law	8/11/67
	4	Massachusetts	Dept. Pub. Health	Sect. 51, Ch 111	10/7/70
	5	New York	Dept. Labor	Code Rule 50	8/1/72
	6	Pennsylvania	Dept. Envir. Resour.	Ch 203, Title 25, Part 1	11/1/71
	7	Texas	Dept. Health	Radiation Control Act— Parts 50, 60, 70	7/2/74
States with existing regulations—or voluntary regulations with no registration requirement	8	Missouri	Dept. Health	Existing Ionizing	—
	9	Montana[2]	Dept. Health & Envir. Sciences	Regulation Applies	—
	10	Virginia	Dept. Health	Reg: 92-003	—
	11	Washington	Dept. Lab. & Ind.	Voluntary Program Ch 296-62-WAC	—
States with enabling legislation passed	12	Arizona[2]	—	HB-5	8/11/70
	13	Arkansas[2]	Public Health	Act 460	—
	14	Florida[2]	Div. Health	Ch 501-122	—
	15	Louisiana[2]	Div. Radiation Control	HB-1165	7/31/68
	16	Mississippi[2]	Dept. of Health	HB-499	4/24/64
	17	Oklahoma[2]	Health Dept.	HB-1405	4/14/69
States with existing OSHA—state agreements[3]	18	California	Labor Dept.	OSHA State	—
	19	Colorado			—
	20	Connecticut			—
	21	Minnesota			—
	22	North Carolina			—
States drafting or awaiting passage of regulations[4]	23	Alabama	—	Pending	—
	24	Iowa			—
	25	Maine	Health Eng.		—
	26	Michigan			—
	27	Nebraska			—
	28	New Hampshire			—
	29	New Mexico	Envir. Impvt. Agency		—
	30	Oregon	Health Division		—
	31	Wyoming	Div. Health		—

[1] Adapted from Reference 10.
[2] New regulations now being drafted or pending passage.
[3] Not covered by other standards at state level.
[4] Enabling legislation not passed.

Five states (18–22 of Table 10-4) have agreements with OSHA, and will at the state level enforce this comprehensive laser user standard. These same states also presently enforce regulations on construction lasers (e.g., "Safety and Health Regulations for Construction"[7]).

Finally, nine additional states (23–31 of Table 10-4) have already developed preliminary drafts of a laser code or are awaiting passage of regulations.

Comparison of Existing State Laser Safety Standards

Of the first eleven states of Table 10-4 that have some existing laser regulations, a mixed picture emerges. Table 10-5 summarizes for these eleven states differences in registration requirements, basis of exposure criteria, if any, and specific requirements for controls and warning signs.

From Table 10-5, it is seen that the regulations for Georgia, Illinois and Pennsylvania only require registration of lasers by the owner with the authorized state agency.

A wide variation exists in the exposure levels for the states of Alaska, Massachusetts, Montana, New York, Texas and Washington.

Montana and Washington have used verbatim the initial 1969 Threshold Limit Values (TLV's) of the American Conference of Governmental Industrial Hygienists (ACGIH),[2] which are now outdated. Those TLV's are applicable to Q-switched and non–Q-switched pulsed ruby lasers. They are more restrictive than the current (1973) ACGIH TLV's for the same type laser.

New York and Alaska have independently set their own exposure criteria, and they are similar to the 1969 ACGIH TLV's but with changes to account for added laser wavelengths and types. The exposure levels are more restrictive in some cases than the revised 1973 ACGIH criteria.

No specific exposure level is given in the Massachusetts standard. Instead it is recommended that reference be made to the ACGIH or ANSI standards for exposure criteria. It does, however, provide exposure guidelines, which are a combination of the 1969 ACGIH values for the visible lasers and the ultraviolet exposure levels from the 1940 American Medical Association's Council on Physical Medicine.

Table 10-5. Comparison of Existing State Laser Safety Standards.[1]

No.	State	Code (Date)	Registration Required	Exposure Criteria Basis	Controls Specified	Warning Signs Specified
1	Alaska	Laser Performance Standard Title 18 Art 7, 8	Yes	Modified 1969 ACGIH TLV's	Yes	Yes
2	Georgia	Ch 270-5-27 (9/1/71)	Yes	Registration Law Only		
3	Illinois	Laser Registration Law (8/1/67)	Yes	Registration Law Only		
4	Massachusetts	Sect. 51, Ch 111 (10/7/70)	Yes	Recommendation only: modified 1969 ACGIH TLV's for visible; also 1940 AMA for UV; also values for IR[4]		No
5	Missouri	Based on existing ionizing regulation	No	No specific Laser Requirements		
6	Montana	Reg 92-003	No	Use 1969 ACGIH TLV's	As specified by ACGIH	
7	New York	Code Rule 50 (8/1/72)	Yes	Modified 1969 ACGIH TLV's	Yes[3]	Yes
8	Pennsylvania	Ch 203-Title 25 Part 1 (11/1/71)	Yes	Registration Law Only		
9	Texas	Radiation Control Act parts 50, 60, 70, (6/2/74)	Yes	Based on ANSI/HEW[2]	Yes	Yes
10	Virginia			Voluntary Laser Program		
11	Washington	Ch 296-62 WAC	No	Use 1969 ACGIH TLV's	As specified by ACGIH	

[1] Adapted from Reference 10.
[2] HEW: Health, Education and Welfare, Food and Drug Administration then proposed Laser Products Performance Standard, Federal Register, Dec. 10, 1973, Vol. 38, No. 236 Part III.
[3] New York requires laser operator certification for all mobile lasers.
[4] AMA: American Medical Association, 1948 Council on Physical Medicine (*JAMA*, 137, 1600, 194).

Table 10-6. Interpretation of New York State Code Rule 50, "Lasers."

Section No.	Description

I. APPLICABILITY

50.2 (a) A. General: Code Rule 50 applies to every person subject to the jurisdiction of the N.Y. State Labor Law who,

    (1) transfers   (5) installs
    (2) receives   (6) tests
    (3) possesses   (7) services
    (4) uses

a laser that may result in the exposure of such person to laser radiation and other hazards associated with lasers.

B. Exemptions: Code Rule 50 is not applicable to:

50.2 (a) (1) Persons subject to the regulations of the
    (a) N.Y. State Department of Health.
    (b) N.Y. City Department of Health.

50.3 (a) (2) Lasers during storage, shipment or sale if they are not capable of emitting laser radiation. However, labeling requirements apply.

50.3 (b) (3) Lasers that cannot emit radiation $>1 \times 10^{-7}$J $\cdot$ cm^{-2} or 1×10^{-5}W $\cdot$ cm^{-2}, measured 10 cm from exterior surface of laser. This exemption not applicable to testing or servicing.

C. Qualifications: This rule does not limit the use of lasers in healing

50.2 (b) (1) Humans when done by or under the supervision of a N.Y. State licensed physician.

50.2 (c) (2) Animals when done by or under the supervision of a N.Y. State licensed veterinarian.

Section 213 of Labor Law D. Violations/Penalties: Any person or corporation officer who knowingly violates Rule 50 or any other industrial code is guilty of a misdemeaner.

50.2 (d) E. Federal Statutes: Nothing in Rule 50 affects the requirements placed upon manufacturers, importers or users of lasers by federal statutes and applicable in N.Y. State.

II. LASER CATEGORIES

A. Exempt Lasers

50.3 (b) (1) Cannot emit radiation $>1 \times 10^{-7}$J $\cdot$ cm^{-2} or 1×10^{-5}W $\cdot$ cm^{-2}, measured 10 cm from exterior surface of laser. Not exempt for testing or servicing of such lasers during production or repair.

50.3 (a) (2) Not capable of emitting radiation during storage, shipment or sale, except that labeling requirements apply.

B. Approved Low-Intensity Lasers

50.23, Table 1 of (1) Low-intensity lasers are capable of energy or power densities:
    (a) $>1 \times 10^{-7}$J $\cdot$ cm^{-2} (Q-switch), $>1 \times 10^{-6}$J $\cdot$ cm^{-2} (non

Table 10-6. (*Continued*)

Section No.	Description
Appendix B	Q-switch) or $>1 \times 10^{-5} W \cdot cm^{-2}$ (CW or pulse width >0.1 sec) for $\lambda = 400$–$1,400$ nm. $>1 \times 10^{-3} J \cdot cm^{-2}$ (Q-switch), $>1 \times 10^{-2} J \cdot cm^{-2}$ (non–Q-switch) or $1 \times 10^{-2} W \cdot cm^{-2}$ (CW or pulse width >0.1 sec) for $\lambda > 1,400$ nm.
50.23, Table 3 of Appendix B	(b) $\leqq 0.1$ J $\cdot cm^{-2}$ (Q-switch), $\leqq 1.0$ J $\cdot cm^{-2}$ (non–Q-switch) or $\leqq 3.0$ W $\cdot cm^{-2}$ (CW or pulse width > 0.1 sec). *Note:* No wavelength dependency indicated for Table 3 low-intensity limits.
50.7 (a)	(2) Any low-intensity laser used or operated in N.Y. State shall be approved.
50.7 (a), (b) and 50.8	(3) Exception: Research and development low-intensity lasers do not require approval, but must be registered and must comply with other applicable provisions of Rule 50.
50.7 (b) and 50.8	(4) Approved low-intensity lasers are exempt from registration.
—	(5) Approval is usually obtained by the laser user, manufacturer or importer.
50.7 and 114.2	(6) Approval Procedure: For approval, use Form BSA-3 (Figure 10-2). *Note:* No provision is stipulated in Rule 50 for approval to use high-intensity lasers.
	C. Registered Lasers/Laser Installations (1) Types Requiring Registration
50.7 (b) and 50.8	(a) Low-intensity lasers used exclusively in research and development.
50.23, Table 3 of Appendix B	(b) All high-intensity lasers, e.g., where power or energy density: >0.1 J $\cdot cm^{-2}$ (Q-switch), >1.0 J $\cdot cm^{-2}$ (non-Q-switch) or >3.0 W $\cdot cm^{-2}$ (CW or pulse width >0.1 sec).
50.8 and 50.4 (t)	(c) Laser Installations [i.e., any location (building, structure or portion therof), other than a construction site, where one or more lasers are used or operated for a period exceeding 30 days].
50.8	(i) Exception: Registration not required if approved low-intensity lasers are exclusively used.
50.8 and 50.4 (aa)	(d) Mobile Lasers (i.e., a laser used or operated outside a laser installation)
50.8	(i) Exception: Registration not required if approved low-intensity lasers are exclusively used.
—	(ii) Usage Requirement: User must file Form IH-331 (Figure 10-3), "Notice of Intent To Use Mobile Laser at Temporary Job Site(s)," prior to using laser on job site.
50.8	(2) Timing of Registration (a) Laser installations and mobile lasers in use on or before

Table 10-6. (Continued)

Section No.	Description
	8/1/72 should have been registered within 90 days after 8/1/72.
	(b) After the 8/1/72 date, registration shall be made prior to receipt or assembly of laser.
—	(3) Registration Procedure
	(a) For registration, use Form IH-313 (Figure 10-4), "Registration of Laser Installation and Mobile Lasers."
	(b) Registration is not completed until stamped copy of completed IH-313 is received from N.Y. State Division of Industrial Hygiene (DIH).

III. EMPLOYER RESPONSIBILITIES

Section No.	Description
50.5	A. Compliance with provisions of Rule 50 relating to laser radiation and other associated hazards:
	(1) Shall not permit any individual to be exposed to laser radiation above the limits of:
50.23	(a) Table 1 (of Appendix B) for cornea
50.23	(b) Table 2 (of Appendix B) for skin
	unless adequately protected.
	(2) Shall not permit any individual to be exposed to associated laser hazards unless adequately protected.
—	B. Obtains approval for usage or operation of low-intensity lasers (see Section II. B of Table 10-6).
	C. Registers all laser installations and mobile lasers [see Section II C. (1) of Table 10-6].
50.12 (a)	D. Shall instruct and advise every individual employed in or lawfully frequenting a laser radiation area regarding
50.12 (a) (1)	(1) The presence of a laser
50.12 (a) (2)	(2) Potential hazards associated with the use of laser.
50.12 (a) (2)	(3) Precautions and procedures necessary to minimize exposure to radiation from the laser.
50.12 (a) (3)	(4) Applicable provisions of Rule 50 for protection from laser radiation.
50.11	E. Designates a laser safety officer.
50.9 (a)	F. Permits only individuals with a valid certificate of competence to operate mobile lasers.
	Exceptions: (1) Professional engineers and land surveyors licensed to practice in N.Y. State.
	(2) Operators of mobile lasers used exclusively in research and development.
50.12 (b) (1) (2)	G. Provides protective equipment for individuals frequenting a laser radiation area.
50.4 (t)	H. The confines of a laser installation shall be designated by the owner of such installation.
50.4 (uu)	I. Surveys (e.g., an evaluation of the laser radiation hazards during production, use, disposal, servicing or presence of any laser):

Table 10-6. (*Continued*)

Section No.	Description
50.16 (a)	The following surveys are required for registered lasers:
50.16 (a) (1)	(1) Survey of output power or energy density of each registered mobile laser, prior to initial use.
50.16 (a) (2)	(2) Survey of all laser protective devices, including safety interlocks, prior to initial use.
50.16 (a) (3)	(3) Visual inspection for optical defects of all safety eyewear at least once every six months. Also verification that optical density and type filter are appropriate for laser in use.
50.17 (a)	J. Safeguards laser against unauthorized use.
50.17 (b)	K. Properly disposes of high-intensity lasers.
	(1) Renders laser permanently inoperative, or
	(2) Transfers laser to other party who has obtained required registration.
50.13 (j)	L. Provide at least 30 lumens per square foot of illumination in a laser radiation area, except where conditions of laser operation require lower ambient illumination.
50.19	M. Maintenance of Necessary Records: The following records must be kept by owners of registered lasers and shall be made available for examination by authorized N.Y. State officials.
50.19 (a)	(1) Results of each required initial laser output check and system interlock check.
50.19 (b)	(2) Transfer, receipt and disposal of any registered laser.
50.19 (c) and 50.16 (c)	(3) Annual inventory of all lasers in any laser installation and all mobile lasers.
50.19 (c) and 50.16 (a) (3)	(4) Safety eyewear checks.
50.19 (c) and 50.16 (b)	(5) Instruments used and calibration data for laser measurements made for compliance with Rule 50.
50.20	N. Reports
	(1) To N.Y. State Industrial Commissioner
50.20 (a) (1)	(a) Any theft or loss of intact laser reported immediately.
50.20 (a) (2)	(b) Written report in seven days of any injury caused by laser or associated equipment.
50.20 (b)	(2) To physician: in case of reported exposure to laser radiation, all pertinent data relating to such exposure shall be made available to any physician authorized by the exposed individual.
50.21	O. Inspection and Tests: The employer permits authorized N.Y. State officials to:
50.21 (a) (1)	(1) Inspect the laser, laser radiation area and laser installation or premises where such laser is located, possessed, stored or used.
50.21 (a) (2)	(2) Inspect all records and reports as per Items M and N above.
50.21 (b)	(3) Conduct tests deemed necessary.
50.21 (c)	*Note:* If any inspection or test by authorized state personnel may compromise national security, a statement

Table 10-6. (*Continued*)

Section No.	Description
	and suitable report from the laser safety officer concerning proper operation of the laser and safety precautions is acceptable.
50.18	P. Provides protection against any associated hazards (e.g., air contaminants, electric shock, etc.) that may be present during the laser operation.

IV. EMPLOYEE RESPONSIBILITIES

50.6	A. Complies with all provisions of Rule 50 relating to his personal conduct.
50.6	B. Properly uses the protective equipment provided for his protection.
50.9 (a)	C. Does not operate any mobile laser unless the individual holds a valid certificate of competence. Exceptions: (1) Professional engineers and land surveyors licensed to practice in the State of New York. (2) Operators of mobile lasers used exclusively in research and development.
50.10 (a) (b)	D. Shall not operate a laser in a manner so as to cause persons to be exposed to radiation levels in excess of (1) Table 1 (of Appendix B) for cornea exposure. (2) Table 2 (of Appendix B) for skin exposure.
50.10 (c) 50.10 (c) and 50.4 (u)	E. Shall not operate a high intensity laser unless: (1) It is placed in shielded and interlocked enclosures, or (2) Operated only in laser radiation areas (i.e., any area containing one or more high- or low-intensity lasers, the access to which is controlled for the purpose of protecting individuals from exposure to laser radiation). Exceptions: (i) Need not enclose, interlock or operate laser in a laser radiation area if Commissioner is notified 48 hours before planned operation, using Form IH 331 (Figure 10-3). Laser safety officer or his designate shall be continuously present when laser is operating. (ii) When servicing or repairing high-intensity lasers, a laser safety officer or his designate shall be present, but the Commissioner need not be notified.
50.13 (a)	F. Does not look directly into primary or specularly reflected laser beam when intensities exceed values of Table 1 (of Appendix B), unless approved eye protection in compliance with Rule 50 is used.
50.13 (f)	G. Laser beam paths shall be cleared of all material that may cause beam scatter in uncontrolled area.

Table 10-6. (*Continued*)

Section No.	Description
50.13 (g)	H. All unshielded lasers exceeding maximum permissible exposure limits shall be capped or otherwise effectively terminated when not in operation.
50.16 (a) (2)	I. Prior to initial laser use, makes certain that all laser protective devices including safety interlocks are in good working order and have been properly installed.
50.17 (a)	J. Safeguards laser against unauthorized use.
50.18	K. Ensures that adequate protection is taken against associated hazards (e.g., cryogenic coolants, ionizing radiation, etc.) that may be present during the laser operation.

V. LASER SAFETY OFFICER (LSO) RESPONSIBILITIES

50.11	A. Establishes and administers laser radiation safety program that is in compliance with Rule 50. *Note:* Each company establishes specific requirements of its laser safety program for its particular laser use; however, the Division of Industrial Hygiene requests that a safety program outline be forwarded to it.
50.11, 50.9 (b) (2) and 50.9 (a)	B. For a high-intensity mobile laser operation, the LSO does not require a Class B certificate of competence or equivalent, but if so qualified may be the operator of such a laser.
50.11	C. The LSO at a given site or location using low-intensity lasers may be (but is not required to be) a certified Class A mobile laser operator.
50.11	D. Appoints, trains and instructs alternate(s) in laser safety techniques, when LSO does not personally supervise the safety aspects of laser operation. *Note:* Training requirements are not specified; however, the Division of Industrial Hygiene will review the training material during a plant survey.
—	E. Is required to sign Form IH-313 (Figure 10-4) for registration of laser installation and mobile lasers.
—	F. Will ordinarily sign and attest to the qualifications of applicant applying for mobile laser certificate of competence, Form DOSH-275 (Figure 10-5), "Application for Certificate of Competence for Operator of Mobile Laser."
50.10 (c) (1)	G. LSO or designee must be present if high-intensity laser is operated without enclosure or interlocks in service, or not in a laser radiation area.
50.10 (c) (2)	H. LSO or designee must be present when repairing or servicing high intensity laser during any period of operation without shielding or without interlocks in service.
50.21 (b)	I. Assists and advises N.Y. State personnel during conduct of laser tests deemed necessary by them.
50.21 (c)	J. Reports to Commissioner (Division of Industrial Hygiene) that

Table 10-6. *(Continued)*

Section No.	Description
	proper operation of laser and safety procedures are in force if any inspection by state personnel would compromise national security.
50.14 (b)	K. When the laser is operated for eight hours or less, the normal requirements for signs and control device equipments for a laser radiation area are waived. However, the LSO or his designee must be constantly in attendance and take all precautions to prevent exposure above maximum permissible radiation limits, and to prevent injury due to associated laser hazards.
50.12 (b) (2)	L. Protective devices such as protective gloves, clothing and shields are to be used by individuals as determined by the LSO.

VI. OPERATORS OF MOBILE LASERS

Section No.	Description
50.4 (aa)	A. A laser used or operated outside a laser installation is a mobile laser.
50.9 (a)	B. Mobile laser operator must have appropriate certificate of competence to operate a laser.
50.9 (b) (1)	(1) Class A certificate—low-intensity mobile laser
50.9 (b) (2)	(2) Class B certificate—low- and high-intensity mobile laser.
	Exceptions: (1) N.Y. State licensed professional engineers or surveyors.
	(2) Operators of mobile lasers used exclusively in research and development.
	C. Requirements
50.9 (e) (1)	(1) Must be at least 18 years of age.
50.9 (d)	(2) Must have no uncontrolled physical handicap or illness, such as epilepsy, heart disease or uncorrected defect in vision or hearing.
	(3) Experience
50.9 (e) (2)	(a) One year of practical experience in the operation of a laser is required, including a knowledge of laser safety precautions.
50.9 (e) (2)	(b) One-year experience requirement may be waived if evidence submitted of successful completion of a laser operator training course.
50.9 (g)	(4) Examination
	(a) New applicant for certificate of competence must pass examination.
	(b) Applicant for renewal of certificate may not require examination.
	Exception: Certificate of competence or renewal could be obtained if applicant successfully completed a four-year course of study at a college or university, including laser theory and laser safety.
50.9 (c)	(5) Must complete Form DOSH-275 (Figure 10-5), "Application

Table 10-6. (*Continued*)

Section No.	Description
	for Certificate of Competence for Operator of Mobile "Laser," and submit photographs.
50.9 (i)	D. Certificate of competence is valid for three years from date of issue, unless suspended, revoked or extended by Commissioner.
50 (j)	E. Mobile laser operator must carry certificate when working with or operating mobile laser, and must produce it for authorized inspector.
50.9 (k)	F. Mobile laser operator must renew certificate not more than six months or less than three months prior to expiration date.
	G. Suspension or Revocation of Certificate of Competence
50.9 (1) (2)	(1) The Commissioner may suspend or revoke a certificate for just cause after:
	(a) Notice to interested parties.
	(b) Hearing before (laser) examining board.
50.9 (1) (4)	(c) Receipt of recommendation from examining board.
50.9 (1) (1)	(2) Just cause for suspension or revocation may be lack of proper competence, judgment or ability with respect to mobile laser operation, or for other good cause.
50.9 (1) (2)	H. Refusal to Renew Certificate of Competence: The Commissioner may refuse to renew a certificate of competence, but only after:
	(a) Hearing before (laser) examining board.
	(b) Notice of hearing sent to interested parties.
50.9 (1) (4)	(c) Receipt of recommendation from examining board.
50.9 (1) (3)	I. Denial of Certificate of Competence: After denial by Commissioner, individual may by written request to Commissioner within 30 days of denial, have a hearing before the (laser) examining board.
50.9 (1) (4)	J. Hearings before the (laser) examining board in connection with suspension, revocation, refusal to review or denial of a certificate of competence, must result in the examining board's making recommendations to the Commissioner within three days after conclusion of such hearings, written notice of Commissioner's decision with reasons shall be forwarded promptly to all interested parties.
50.9 (f)	VII. (LASER) EXAMINING BOARD
50.9 (f) (1)	A. The Commissioner appoints examining board members.
	(1) Board consists of minimum of three members.
	(2) At least one member must hold a valid class B certificate of competence to operate a mobile laser.
	B. Duties of (Laser) Examining Board
50.9 (f) (2) (i)	(1) Examines applicants for a certificate of competence on their qualifications (experience and competence).
	(2) Holds necessary hearings:
50.9 (f) (2) (ii)	(a) Appeals following denials of certificate.

Table 10-6. *(Continued)*

Section No.	Description
50.9 (f) (2) (iii)	(b) Prior to determination by Commissioner to suspend or revoke certificates or to refuse to issue renewal of certificate.
50.9 (f) (2) (iv)	(3) Reports findings and recommendations of hearings to Commissioner.
50.9 (f) (3)	C. The Commissioner may designate panels to conduct examinations and hearings on behalf of the examining board.

VIII. LASER RADIATION AREA HAZARD CONTROL MEASURES AND PRECAUTIONS

50.4 (u)	Definition of Laser Radiation Area: Any area containing one or more high- or low-intensity lasers, the access to which is controlled for the purpose of protecting individuals from exposure to laser radiation.
50.13 (b)	A. When necessary, the laser beam shall be terminated by a material that is nonreflecting and fire retardant.
50.13 (b)	B. If the laser beam heats, melts and splatters target material, precautions should be taken to prevent fires arising from flying particles.
50.13 (b)	C. For unavoidable reflections, precautions shall be taken to prevent scatter of laser radiation into uncontrolled areas in excess of maximum permissible exposure limits.
50.13 (c)	D. The area shall be cleared of all individuals for a reasonable distance on all sides of the anticipated laser beam path, where operation can lead to accidental exposure above maximum permissible limits, unless adequate protection in compliance with Rule 50 can be provided.
50.13 (d)	E. Safety eyewear shall provide the required attenuation or optical density (O.D.) for the power and at the wavelength of the laser in use.
50.13 (f)	F. Laser beam paths shall be cleared of all material that may cause scatter in an uncontrolled area.
50.13 (g)	G. All unshielded lasers exceeding maximum permissible exposure limits shall be capped or otherwise effectively terminated when not in operation.
50.13 (i)	H. A closed laser installation with an entry shall have at least one exit door that can be easily opened from the inside.
50.13 (j)	I. General illumination shall be 30 lumens per ft^2, unless laser operation requires lower illumination.
	J. Laser Radiation Area Signs
50.14 (a)	1. The laser radiation sunburst symbol shall be used to designate laser radiation areas (as well as lasers).
50.14 (b)	2. "CAUTION" or "DANGER" and "LASER RADIATION AREA" signs shall be conspicuously placed at the entrance to and inside of each laser radiation area.

Table 10-6. (*Continued*)

Section No.	Description
	Exceptions: (1) No signs are required if the laser operates eight hours or less and the LSO or his designee is in constant attendance. All precautions must be taken to prevent exposure above maximum permissible radiation limits and to prevent injury due to associated laser hazards.
50.3 (b)	(2) Lasers that are exempt (see Section II A of Table 10-6).
50.18	K. Protection shall be provided against associated laser hazards.
50.12 (a)	L. All persons employed in or frequenting a laser radiation area shall be informed about:
	(1) The presence of a laser.
	(2) The potential hazards associated with the use of a laser.
	(3) The precautions and procedures necessary to minimize hazards.

IX. LASER HAZARD ENGINEERING CONTROL MEASURES

50.13 (h)	A. Interlocks used on enclosures for high-intensity lasers shall not be used to acutate the laser.
50.13 (k)	B. Electronic firing systems for pulsed lasers shall be designed to prevent accidental discharges of stored charges.
50.13 (1)	C. "Fail-safe" safety circuitry shall be used with lasers wherever necessary.
50.10 (d)	D. The laser radiation exposure values produced by a scanned laser beam shall be determined.
50.16 (b)	E. Instruments used for laser measurements shall be of the appropriate type and properly calibrated.

X. PERSONAL PROTECTION

50.12 (a)	A. General: Each individual frequenting a laser radiation area should be apprised of:
	(1) The presence of a laser
	(2) The potential hazards associated with the use of a laser.
	(3) The precautions and procedures necessary to minimize hazards.
	(4) Applicable provisions of Rule 50 for the protection of individuals exposed to laser radiation.
50.12 (b) (1)	B. Approved Safety Eyewear
	(1) Is to be used for unshielded laser operations under conditions where accidental exposure to radiation above maximum permissible levels is possible.

Table 10-6. (*Continued*)

Section No.	Description
50.12 (b) (1) (i)	(2) The optical density of the eyewear should reduce radiation to the cornea of the eye to safe levels shown in Table 4 (of Appendix B).
50.12 (b) (1) (ii)	(3) Safety eyewear shall be designed and tested to ensure that it retains its protective properties during use.
50.12 (b) (1) (iii)	(4) Eyewear shall be legibly labeled with the optical density and wavelength.
50.12 (b) (1) (iv)	(5) An individual requiring prescription lenses shall be provided with approved eye protection or with approved safety eyewear designed to fit over his regular prescription lenses.
50.13 (e)	(6) Safety eyewear that has been exposed to high-intensity laser radiation shall not be reused until it has been reevaluated for shattering and effectiveness of its attenuation.
50.16 (a) (3)	(7) Eyewear must be inspected at least every six months to assure that optical density is appropriate to the laser in use, and to visually determine that no optical defects exist.
50.16 (a) (3)	(8) Deteriorated or defective eyewear shall be withdrawn, repaired or discarded.
–	(9) Use Form BSA-3 (Figure 10-2) when applying for approval of safety eyewear.
50.12 (b) (1)	(10) Eyewear shall be provided by the employer to individuals in a laser area when required.
50.12 (b) (2)	C. Other Personal Protective Devices
	(1) Protective gloves, clothing and shields shall be provided by the owner.
	(2) Protective devices are to be used by individuals as determined by the laser safety officer.

XI. LASER LABELS

50.15 (b) and 50.15 (a)	A. All lasers shall be legibly marked in $\frac{1}{8}$-inch-high or greater lettering, with the following information:
50.15 (a) (1)	(1) Laser radiation sunburst symbol in red.
50.15 (a) (2)	(2) The words "CAUTION" or "DANGER" and "LASER" in black letters.
50.15 (a) (3)	(3) The wavelength or wavelengths of the emitted laser radiation.
50.15 (a) (4)	(4) The maximum output power or energy density of the emitted radiation.
50.15 (a) (5)	(5) The divergence of the emitted radiation in its lowest-order tranverse mode.
	B. Exceptions
50.3 (b)	(1) Lasers that are exempt from Rule 50, e.g., those lasers that cannot emit radiation $>1 \times 10^{-7}J \cdot cm^{-2}$ or $>1 \times 10^{-5}W \cdot cm^{-2}$ when measured 10 cm from the exterior surface of such laser.

Table 10-6. (*Continued*)

Section No.	Description
50.15 (b) (1)	(2) If $\frac{1}{8}$-inch-high or greater lettering is not practical because of laser size or location, an alternate manner of marking is acceptable.
50.15 (b) (2)	(3) Lasers used exclusively in research and development need not specify wavelength, maximum power output or energy density or the beam divergence, where such parameters may be unknown or difficult to obtain.

XII. ASSOCIATED LASER HAZARDS

50.18	Protection against associated laser hazards shall be provided. Such associated hazards include but are not limited to the following:

50.18 (a)	A. Air Contaminants
	(1) Can include vaporized target materials and toxic gases, vapors and fumes.
	(2) Such dangerous air contaminants shall be removed or controlled in accordance with:
	   (a) Industrial Code Part (Rule No.) 12 relating to "Control of Air Contaminants" and
	   (b) Industrial Code Part (Rule No.) 18 relating to "Exhaust Systems."
	(3) Adequate ventilation shall be maintained in each laser installation.

50.18 (b)	B. Ultraviolet Radiation (1) Levels $>0.5 \times 10^{-6}\text{W} \cdot \text{cm}^{-2}$ for seven hours must be avoided. (2) Continuous exposure to levels $>0.1 \times 10^{-6}\text{W} \cdot \text{cm}^{-2}$ must be avoided. (3) Special care must be taken to protect individuals from ultraviolet radiation when quartz tubes are used in a laser system, since quartz transmits such radiation very efficiently.
50.18 (c)	C. Electrical Hazards: Since a laser and its associated electrical system may present electrical hazards, the design, construction, installation and maintenance must be such as to minimize the possibility of the electrical hazard.
50.18 (d)	D. Cryogenic Coolants: The storage, handling and use of cryogenic coolants associated with lasers shall be such so as to minimize any hazards that such coolants may present.
50.18 (e)	E. Fire Hazards (1) Design, construction, installation, operation and maintenance of laser and laser installation shall be such as to eliminate or reduce fire hazards. (2) The laser radiation area shall be maintained clear of any unnecessary materials in order to minimize any fire hazard.

Table 10-6. (*Continued*)

Section No.	Description
50.18 (f)	F. Explosion Hazard: Since lasers and associated equipment may present an explosion hazard, such equipment shall be protected with explosion shields to prevent injury to individuals from such hazards.
50.18 (g)	G. Ionizing Radiation: If a laser system produces ionizing radiation, it must be in compliance with the provisions of Industrial Code (Rule No.) 38 relating to "Radiation Protection" with respect to protection against ionizing radiation.

Note: Table 10-6 adapted from Reference No. 11.

Virginia supports a voluntary program of laser safety but has not set any guidelines.

Texas has issued a comprehensive standard using as a guide the ANSI Z-136.1 (1973) standard and the then-proposed (and now completed) BRH federal performance standard for laser products. The Texas standard is supposed to be an initial effort toward a "model state" standard.

In summary, it is seen that existing state standards contain many differences. Further, many of the standards are more restrictive than any of the major national standards such as BRH, ANSI or ACGIH.

Interpretation of New York State Industrial Code Rule 50, "Lasers"

New York State Code Rule 50[4] for the safe use of lasers is a fairly complex standard. Table 10-6, adapted from a bulletin prepared by the Education and Safety Committee of the Laser Institute of America,[11], provides an interpretation of the New York standard. As a supplement, and for the purpose of enhancing the usefulness of the Table 10-6 interpretation, samples of the various forms used by the New York State Labor Department for implementing Rule 50 are included as Figures 10-2 to 10-5.

Finally, a copy of Code Rule 50 is included as Appendix B.

APPLICATION NO.

DATE FILED

State of New York   Department of Labor

BOARD OF STANDARDS AND APPEALS

Tower Building
Empire State Plaza
Albany, New York 12223

BOARD USE

APPL. _____ AGENT _____
DISS DIRECTOR _____
F & M _____ CONST. _____
C.I.UP_____ C.I.DOWN _____
SUPER DIST. # _____(2)
OTHER _____

Application is hereby made for GENERAL APPROVAL of a product or device as follows:

APPLICANT

NAME   ADDRESS

                                                                        ZIP CODE

TELEPHONE NO. (INCLUDE AREA CODE)   WILL COLLECT CALLS BE ACCEPTED? ☐ YES ☐ NO

☐ INDIVIDUAL   ☐ PARTNERSHIP   ☐ AGENT APPLYING - ENTER NAME,   ☐ A CORPORATION INCORPORATED IN THE
                                    ADDRESS AND TELEPHONE NUMBER   STATE OF
                                    UNDER REMARKS

DESIGNATION (GIVE DESIGNATION AS YOU DESIRE IT LISTED)

PURPOSE

PRODUCT OR DEVICE CAN BE EXAMINED AT THE BOARD'S OFFICE   HAS A PREVIOUS APPLICATION FOR ITS APPROVAL BEEN FILED WITH
                                                            THIS BOARD?
☐ YES   ☐ NO   ☐ YES   ☐ NO (IF YES GIVE DETAILS UNDER REMARKS)

HAVE ANY COMPLIANCE ORDERS AFFECTING THIS PRODUCT OR   TO YOUR KNOWLEDGE HAS ANY VARIANCE BEEN REQUESTED OR
DEVICE BEEN ISSUED BY THE INDUSTRIAL COMMISSIONER?   GRANTED FOR THE USE OF THIS PRODUCT OR DEVICE?
☐ NO   ☐ YES (GIVE DETAILS UNDER REMARKS)   REQUESTED ☐ YES ☐ NO   GRANTED ☐ YES ☐ NO
                                                        (GIVE DETAILS UNDER REMARKS)

DATA TO BE INITIALLY SUBMITTED   NOTE: ADDITIONAL DATA MAY BE REQUESTED

1. Completed Application Form.
2. Assembly and detail drawing, including schematic diagrams, indicating the general dimensions, size and materials of all component parts.
3. Typewritten description, in duplicate, of the construction, use, operation and safety features of product or device in layman's language. (Sales brochure NOT acceptable.)
4. Advertising literature, photographs, illustrations or other data, where they are helpful in an understanding of the product or device.
5. Copies of report of any tests, engineering analyses, and of any approvals previously granted to the product or device by any other device by any other Agency. Tested by an Independent Laboratory ☐ yes ☐ no
6. A sample device, when requested by the Board.

REMARKS (Attach additional sheets if necessary):

_____   _____
DATE   SIGNATURE OF APPLICANT OR AGENT

                                            _____
                                            PRINT OR TYPE ABOVE NAME

WARNING: THIS APPLICATION WILL BE SUBJECT TO DISAPPROVAL FOR NON-APPEARANCE AT A SCHEDULED HEARING OR IF ADDITIONAL INFORMATION REQUESTED BY THE BOARD IS NOT FURNISHED.

BSA-3 (6-74)   OVER.....

Figure 10-2. Form BSA-3, for approval to use material or device.

GENERAL INFORMATION AND INSTRUCTIONS

A. TYPES OF APPROVALS

1. GENERAL - Approvals which apply not only to the product or device submitted but to all duplicates of same which are produced.

2. SPECIAL - Approvals of one product or device, usually for use at a single locality, which approval does not apply to duplicates thereof.

NOTE: Either of the above types of approvals may be:
"REQUIRED" A product or device expressly required to be approved by the Labor Law or Industrial Code; or
"VOLUNTARY" A product or device which, although not expressly required to be approved, are or may be used in any industry, trade, occupation, process or establishment subject to the safety provisions of the Labor Law.

B. ACTION BY THE BOARD

1. All costs for setting up equipment for the proper demonstration or testing of product or device must be borne by the applicant.

2. When deemed necessary, an inspection and demonstration of the product or device will be witnessed by New York State Personnel. If such inspection and demonstration is outside the State of New York, the cost of all travel and subsistence expenses are required to be reimbursed by the applicant.

3. When requested, the applicant will be required to submit from 150 to 450 copies of drawings or illustrations in such a size and containing such data as the Board may direct.

SAMPLES

Samples submitted for examination or testing may be retained and made part of the Board's records herein, or may be returned to the Applicant at his expense, as the Board may determine. The sample may be tested as the Board may require and the Board shall not be liable for test damage.

Figure 10-2. *(Continued).*

STATE OF NEW YORK
DEPARTMENT OF LABOR
DIVISION OF INDUSTRIAL HYGIENE
80 CENTRE STREET
NEW YORK, N.Y. 10013

NOTICE OF INTENT TO USE MOBILE LASERS AT TEMPORARY JOB SITE(S)

(Submit Three Copies to Address Below)

TO: State of New York - Department of Labor   Date:
     Division of Industrial Hygiene
     Radiological Health Unit
     80 Centre Street
     New York, N.Y. 10013

1. LASER REGISTRATION NO.

2. NAME OF FIRM

3. ADDRESS OF FIRM (INCLUDING ZIP CODE)

4. ADDRESS(ES) OF JOB LOCATION(S)

(1)

(2)

(3)

(4)

(5)

	5. WORK SCHEDULE		6. LASER(S)		7. RESPONSIBLE PERSONNEL AT SITE
	A. DATE	B. TIME	A. MANUFACTURER	B. MODEL NO.	CERTIFIED MOBILE LASER OPERATOR
(1)					
(2)					
(3)					
(4)					
(5)					

8. SIGNATURE OF OFFICER OF FIRM   9. TITLE

FOR OFFICE USE ONLY

10. FIELD INSPECTION ☐ YES: ☐ NO:	11. DATE	12. IN COMPLIANCE		13. VIOLATIONS
LOCATION NUMBER		YES	NO	

IH - 331 (8-72)

Figure 10-3. Form IH-331, notice of intent to use mobile lasers at temporary job site.

STATE OF NEW YORK   DEPARTMENT OF LABOR

DIVISION OF SAFETY AND HEALTH
Two World Trade Center, New York, NY. 10047

Registration No. LSR

Date Registered

Please Submit:
Form in Triplicate
Check or Money Order

REGISTRATION OF LASER INSTALLATIONS AND MOBILE LASERS
(SEE REVERSE FOR INSTRUCTIONS)

1. NAME OF OWNER (FIRM OR LESSEE)

2. ADDRESS (IN FULL) WHERE LASER IS USED OR STORED   COUNTY

2A. ADDRESS OF FIRM IF DIFFERENT FROM ABOVE

3. CONFINES OF INSTALLATION

4. INDUSTRY   5. TOTAL NUMBER OF WORKERS EXPOSED TO LASER RADIATION

6. a. Are the lasers reported in item 7 a part of a laser installation or mobile laser previously registered with Division of Industrial Hygiene ? ☐ Yes; ☐ No;

   b. If yes, what is our registration number for that installation ? # _____

7. LASER EQUIPMENT (USE ADDITIONAL SHEETS IF NECESSARY)

a. Number Fixed	Mobile	b. Mfg.	c. Model No.	d. Wave Length(s) (nm or μm)	e. Output Power or Eng. Density	f. Beam Dia.	g. Beam Divg.	h. Purpose or Use	i. No. of Workers Exp.

8. a. NAME, TITLE AND BUSINESS ADDRESS OF LASER SAFETY OFFICER (See Definition on Reverse)

   b. QUALIFICATIONS: _____

DATE   SIGNATURE OF LASER SAFETY OFFICER

DOSH–280.1 (5–77)

Figure 10-4. Form IH-313, registration of laser installation and mobile lasers.

INSTRUCTIONS FOR COMPLETION OF DOSH–280.1

(Numbers Correspond to Items on the Front)

1.   Name of Owner – Industrial Code Rule No. 50, "Lasers" defines owner as: "Any person conducting the business or activities carried on within a laser installation or with a mobile laser and having by law the administrative control of a laser, whether as owner, lessee, contractor or otherwise".

2.   Address in Full Where Laser is Used or Stored – Number, Street, Village, Town or City where laser is used or stored. In rural areas give road intersections, highway number, etc. Give county in all cases.

3.   Confines of Installation –

For Laser Installation: Industrial Code Rule No. 50 defines Laser Installation as: "Any location where, for a period of mor·· than 30 days, one or more lasers are used or operated. The confines of a laser installation shall be designated by the owner of such installation. An entire building or other structure, a part thereof or a plant may be designated as a laser installation. For the purposes of this Part (rule), a construction site shall not be considered a laser installation."

For Mobile Lasers: Write "Mobile Laser" and indicate room or area where laser is stored; submit form DOSH–280.2, "Notice of Intent to use Mobile Laser at Temporary Job Site", when laser is used at temporary job site.

4.   Industry – The principal product or type of activity at this location e.g., ordnance manufacturer, printing, electrical machinery mfg., communications, banking, etc. If you know the Federal Government's Standard Industrial Classification for your plant, please enter that Major Group Number.

5.   Total Number of Workers Exposed to Laser Radiation – The total number of workers exposed.

7.   Laser Equipment – In case of lasers used in research and development where the wave length, output power or energy density, beam diameter or beam divergence are unknown, the word "Unknown" should be entered in 7d, 7e, 7f and 7g.

8.   "LASER SAFETY OFFICER"– An individual, designated at a particular laser installation or for a particular mobile laser, who is qualified by training and experience in the occupational and public health aspects of lasers to evaluate the radiation hazards of such laser installation or mobile laser and who is qualifed to establish and administer a laser radiation protection program for such laser installation or mobile laser.

LASER REGISTRATION FEE

A check or money order for $300.00, payable to the Industrial Commissioner, must accompany each registration of a laser installation or mobile laser. This registration is valid for three years, after which a new fee is required.

A stamped copy of your registration form will be returned to you, which should be retained as proof of your registration and payment of fee.

Figure 10-4. (*Continued*).

DATE RECEIVED	FEE	PHOTOS		CERT. NO.	EXPIRES	CLASS

DO NOT WRITE ABOVE THIS LINE

Submit in Duplicate
with
Two Photographs

STATE OF NEW YORK – DEPARTMENT OF LABOR
Division of Safety and Health
Two World Trade Center
New York, N.Y. 10047

APPLICATION FOR CERTIFICATE OF COMPETENCE FOR OPERATOR OF MOBILE LASER

1. LAST NAME OF APPLICANT	FIRST	MIDDLE	2. SOCIAL SECURITY NUMBER

3. PERMANENT ADDRESS - WHERE YOU ARE CERTAIN OF BEING REACHED

3A. NUMBER AND STREET		3B. COUNTY
3C. CITY, TOWN, VILLAGE	3D. STATE	3E. ZIP CODE

4. PLACE OF RESIDENCE FOR NEXT SIX MONTHS IF OTHER THAN SHOWN IN 3 ABOVE

4A. NUMBER AND STREET		4B. COUNTY
4C. CITY, TOWN, VILLAGE	4D. STATE	4E. ZIP CODE

5. DATE OF BIRTH	6. WEIGHT	7. HEIGHT	8. COLOR EYES	9. COLOR HAIR

10. DO YOU HAVE A PHYSICAL HANDICAP OR ILLNESS, SUCH AS EPILEPSY, HEART DISEASE, OR AN UNCORRECTED DEFECT IN VISION OR HEARING, THAT MIGHT DIMINISH YOUR COMPETENCE IN RELATION TO THE HANDLING OR USE OF LASERS?

☐ NO   ☐ YES ...... IF YES, EXPLAIN BELOW!

11. PLEASE CHECK ONE BOX TO INDICATE THE CATEGORY OF THE CERTIFICATE OF COMPETENCE REQUIRED:

☐ Class A - Certificate - The holder of this certificate may operate any low-intensity mobile laser.

☐ Class B - Certificate - The holder of this certificate may operate any high-intensity or low-intensity mobile laser.

12. ☐ RENEWAL APPLICANT - Check here and sign reverse of form.

☐ FIRST-TIME APPLICANT - Check here and complete ALL of reverse of form.

DOSH-275 (5-77)

Figure 10-5. Form DOSH-275, application for certificate of competence for operator of mobile laser.

13. WORK EXPERIENCE - List Below Your Work Experience in the Handling,Use and Operation of Mobile Lasers (Use Additional Sheet, If Necessary)
   Note: Letters from your employers verifying experience claimed will expedite processing of your application

13a. DATE(S) EMPLOYED IN THE HANDLING, USE AND OPERATION OF MOBILE LASERS	13b. EMPLOYER'S NAME, MAILING ADDRESS AND LASER SAFETY OFFICER	13c. LASERS USED		13d. BRIEFLY DESCRIBE YOUR DUTIES
		MFG.	MODEL NO.	
FROM: TO:	LSO:			
FROM: TO:	LSO:			
FROM: TO:	LSO:			
FROM: TO:	LSO:			
FROM: TO:	LSO:			
FROM: TO:	LSO:			
FROM: TO:	LSO:			

14. LASER TRAINING COURSE(S)

14a. DATE(S)	14b. SPONSOR OF TRAINING COURSE AND COURSE DIRECTOR	14c. LOCATION OF TRAINING	14d. NAME OF INSTRUCTOR	14e. STATEMENT OF COMPLETED COURSE-SENT BY COURSE DIRECTOR	
				YES	NO
FROM: TO:					
FROM: TO:					
FROM: TO:					
FROM: TO:					

15. I hereby certify that all the above statements are true to the best of my knowledge and belief.

_____   _____
(SIGNATURE)   (DATE)

IMPORTANT NOTICE   A fee of $30.00 will be charged at the time your full certificate is
REGARDING FEE: issued. Send no money at this time

Figure 10-5. (*Continued*).

REFERENCES

1. "American National Standard For the Safe Use of Lasers, Z 136.1-1976," American National Standards Institute, New York, 1976.
2. "A Guide for Control of Laser Hazards," American Conference of Governmental Industrial Hygienists, Cincinnati, Ohio, 1973.
3. "Laser Products Performance Standards," *Federal Register*, Vol. 40 No. 148, pp. 32252–32266, U.S. Dept of Health, Education and Welfare, Food and Drug Administration, Bureau of Radiological Health (BRH), Washington, D.C., 31 July 1975.
4. Industrial Code Rule 50, "Lasers," State of New York, Department of Labor, Board of Standards and Appeals, Albany, N.Y., August 1, 1972.
5. "Guide for Submission of Information on Lasers and Products Containing Lasers Pursuant to 21 CFR 1002.10 and 1002.12," U.S. Dept. of Health, Education and Welfare, Public Health Service, Food and Drug Administration, Bureau of Radiological Health, Division of Compliance, Rockville, Maryland, July 1976.
6. "Quality Control Practices For Compliance With the Federal Laser Product Performance Standard," U.S. Department of Health, Education and Welfare, Public Health Service, Food and Drug Administration, Bureau of Radiological Health, Division of Compliance, Rockville, Maryland, March 1976.
7. "Safety and Health Regulations for Construction," Department of Labor, Bureau of Labor Standard, Washington, D.C., April 17, 1971.
8. Sliney, D. H., Franks, J. K., "Safety Rules and Recommendations," *Laser Focus Buyers Guide*, Advanced Technology Publications Inc, Newton, Mass., 1977.
9. "Practice for Occupational and Educational Eye and Face Protection, Z 87.1," American National Standard Institute, New York, 1968.
10. Rockwell, R. J., Jr., "Current Status of State Laser Safety Legislation, Report From the L1A Laser Safety Committee," *Laser Institute of America News*, April 1975.
11. "Interpretation of New York State Industrial Code Rule 50 'Lasers' and Advisory Standards on Low Intensity Lasers and Safety Eyewear," Educational Committee of the Laser Institute of America, Cincinnati, Ohio, October 1974.

11
Laser Safety Program

INTRODUCTION

The key goal of any laser safety program is to prevent individuals from receiving harmful exposures to laser radiation and/or injury from associated potential hazards (e.g., electrical shock, air contaminants and others).

It is the responsibility of the employer (i.e., management) to organize and implement a suitable program for controlling all laser hazards. At least the following broad elements should be included in a laser safety program:

- Clear channels of authority and responsibility for laser hazard evaluation.
- Clear channels of authority and responsibility for controlling laser hazards.
- Suitable education and training of authorized user and supervisory personnel regarding laser hazards.
- Knowledge, understanding and proper utilization of appropriate standards (i.e., regulations) for the assessment and control of laser hazards.
- Use of protective gear if laser control measures still present a potential hazard.

220

- Clear recognition by employee of responsibilities when working with or near lasers.
- Medical surveillance of personnel using lasers.
- Management of suspected or actual laser accidents.

Laser safety programs for different organizations will vary with the laser system application and complexity, magnitude of the laser hazards, experience in dealing with laser hazards and other factors. Nonetheless, general guidelines can be established for the organization of a laser safety program, which can then be suitably altered to reflect the peculiarities of each operation and organization. Such a guide is provided in the section that follows. Included in this discussion will be specific references to documented laser safety program aspects covered by IBM[1] and New York State Industrial Code Rule 50, "Lasers."[2]

Other subjects covered in this chapter are:

- Laser safety training programs
- Audio/visual sources for training programs

GENERAL GUIDELINES FOR ORGANIZATION OF LASER SAFETY PROGRAM

The following elements require consideration in organizing a satisfactory laser safety program:

- Laser safety committee
- Laser safety officer
- Deputy laser safety officer
- Laser system supervisor
- Employee
- Documentation of safety procedures and responsibilites

Modifications, additions and deletions to this list can be instituted to reflect specific needs and requirements.

Laser Safety Committee

Four aspects of a laser safety committee are addressed:

- Need

- Membership
- Responsibilities
- Duration

The need for organizing a laser safety committee is strongly dependent on the following factors:

- Magnitude of the potential laser hazards.
- Complexity of the laser system.
- The nature and/or experience of the organization in dealing with lasers.
- The extent of peripheral disciplines involved in the laser system and their influence in altering the laser hazard.
- Need and degree of concern in complying with public laws governing the safe use of lasers (e.g., New York State Industrial Code Rule 50^2 and others).

As for membership in the laser safety committee, a core requirement is inclusion of individuals well-versed in laser technology, hazard assessment and control, and public regulations. Laser engineers and scientists could fit the bill, particularly in the initial stages of organizing and implementing a laser safety program. Technical management personnel and the designated Laser Safety Officer would complete the hard core. A medical officer might likewise be needed, as would any other key individuals significantly concerned with or affecting the laser hazard.

The fundamental responsibility of the committee is to develop and implement satisfactory safeguards, practices and procedures for controlling laser hazards and thus preventing individuals from sustaining injuries.

As policies and practices dealing with laser safety are developed and implemented, as experience is gained with a particular product or project, and when these considerations, *in toto* tend to become routine, the need to continue the committee generally diminishes, and it may be terminated. The Laser Safety Officer and/or line management for the laser product would handle the daily routine laser safety management.

Laser Safety Officer (LSO)

The LSO should have broad authority for management of the control of laser hazards. He ultimately bears responsibility for assuring that individuals do not receive harmful exposure to laser radiation and/or injury from associated potential hazards. His responsibilities should include at least the following items:

- Serving as a center of contact for all laser safety matters
- Regulations
- Review of planned installations or plans for modified installations
- Approval of environment for laser system operation
- Approval of laser system operation
- Laser operator certification
- Protective equipment usage approval
- Warning systems and signs
- Inspection of laser equipment areas
- Inventory control and records
- Personnel records of authorized or potential laser users
- Management of suspected or actual laser accidents

The LSO should be the center of contact for all laser safety matters, both within the organization and with outside government agencies, companies and individuals. He should provide consultative assistance on laser hazard evaluation and control. Should a laser safety committee exist, membership and attendance at each meeting by the LSO are mandatory. The requirement for an LSO, and the major responsibilities for laser safety, are made quite clear both in N.Y. State Rule 50[2] for the safe use of lasers and in IBM's *Laser Safety-Manager's Reference Manual.*[1] Rule 50 states that the LSO is responsible for establishing and administering a laser safety program, and is the cited authority to assist and advise N.Y. State personnel during conduct of laser tests deemed necessary by them.

Should the laser safety committee approach be deemed unnecessary, it is the LSO who should develop and implement satisfactory rules and procedures for controlling laser hazards. Indeed, according to Reference 1, at IBM the LSO has the responsibility of determining the proper classification of each laser in accord with the applicable standards (i.e., regulations), such as various state standards, BRH standards or voluntary standards such as those of ANSI[3] or ACGIH[4].

The LSO should review planned installations or plans for modified laser installations and give his approval only when he is convinced that the laser hazard controls are satisfactory. IBM mandates that the department manager notify the LSO prior to procurement or construction of laser equipment. Further, notification must likewise be given the LSO when laser operation is discontinued or transferred to another location, within a facility or to another facility.

Approval of the environment for laser system operation should be under the scrutiny of the LSO. Code Rule 50 requires the presence of an LSO for the firing of a high-intensity laser not in a laser radiation area. IBM requires approval of all laser environments before the operation of a laser.

Approval of laser system operation by the LSO should be given only when he is convinced that laser hazard controls are satisfactory. If not so convinced he should have the authority to suspend, restrict or terminate lasing. IBM requires that the LSO be notifed by the department manager prior to initial operation of laser equipment.

Prior to operation of any laser system, the LSO should be informed that the laser operator has been properly certifed. The certification process may be based strictly on internal organization procedures or on government regulations. At IBM, the LSO upon receiving an eye exam report and notice of the employee's having completed laser safety training will sign a certification card. In N.Y. State the LSO will ordinarily attest to the qualifications of the applicant applying for a mobile laser certificate of competence, Form IH-385, "Application for Certificate of Competence for Operation of Mobile Laser."

Where protective equipment is needed for controlling laser hazards, such equipment should be approved by the LSO. Code Rule 50 requires that protective devices such as gloves, clothing and eye shields be used by individuals as determined by the LSO. In regard to eye protection only, IBM requires selection by the LSO but issuance by the company optometrist to assure proper fit.

The LSO must ensure that satisfactory warning systems and signs are placed and/or installed at suitable locations. Chapter 8, Control of Laser Radiation Hazard, details the necessary warning systems and signs.

Complete freedom to inspect all laser equipment areas should be provided to the LSO.

Inventory control of all lasers, along with the necessary records, should be maintained by the LSO. This is particularly important when government regulations require inventory control. Code Rule 50 requires owners of registered lasers to keep records of:

- Annual inventory of all lasers in any laser installation and all mobile lasers.
- Transfer, receipt and disposal of any registered laser.

IBM specifies the LSO as responsible for maintaining a current inventory of all laser devices.

Appropriate personnel records shall be maintained by the LSO:

- Of individuals expected to work with or near nonexempt lasers, as provided by the laser system supervisor.
- For scheduling medical examinations.
- For assuring that scheduled medical examinations have been completed.
- Of suspected or real laser accidents.

IBM requires that all employees assigned to work with lasers have an authorization card signed by the department manager and company physician upon completion of an eye examination. This card must first be forwarded to the LSO.

Upon being informed of a suspected or real laser accident, the LSO must expeditiously investigate the accident and take the necessary follow-up action. Code Rule 50 requires that the N.Y. State Industrial Commisioner receive a written report within seven days of an injury caused by a laser or associated equipment. Further, all pertinent data relating to such exposure must be made available to any physician authorized by the individual believed injured.

Deputy Laser Safety Officer

Selection of a deputy laser safety officer is frequently necessary when multiple laser operations must be supported or when the LSO's presence is not possible. Code Rule 50 stipulates that the LSO appoints, trains and instructs alternate(s) in laser safety techniques when the LSO does not personally supervise the safety aspects of laser operation.

Laser System Supervisor

The supervisor of any laser operation evidently is a key element in any effective laser safety program. The laser system supervisor should be responsible for at least the following items:

- Understanding his responsibilities
- Indoctrination/training
- Control of laser hazards
- Approval of initial and subsequent laser system operation
- Approval of planned installations or plans for modified installations
- Information flow for potential laser users
- Suspected or real laser accidents

In order to implement his responsibilites, the laser system supervisor must understand those responsibilites and the required organizational interface. IBM spells out the responsibilities in their *Laser Safety-Manager's Reference Manual*,[1] which identifies notification requirements, inventory data, employee training requirements, medical examinations, eye protection procedures, environment approval plan and certification requirements. Samples of all relevant forms, labels, signs and certification card are also included in the manual. The document is issued to each manager when he takes a Computer Assisted Instruction (CAI) Laser Safety Course. The course is a requirement for all managers of personnel involved with lasers.

The laser supervisor is responsible for the indoctrination and training on laser hazards and control, for employees under his jurisdiction. IBM requires the departmental manager to ensure that employees under his responsibility partake in required training prior to operating a laser. The training must include the safety aspects of the lasers and the correct operation of the specific unit(s) to be used.

Lasing in the presence of authorized users, other employees, visitors or the public should be permitted by the laser supervisor only when satisfactory control of laser hazards has been provided.

Initial operation of a potentially hazardous laser system under the supervisor's cognizance will only be done after permission has been secured from the LSO. Indeed, this is a specific IBM requirement placed on their management involved with lasers. Depending on the company, subsequent firings of potentially hazardous lasers may require LSO approval. In general, where modifications to a laser system are made that can introduce additional hazards, the supervisor should assure that lasing is not done without prior LSO approval.

Whenever feasible, laser installation or modification plans should be provided to the LSO by the laser supervisor, for approval relative to laser safety considerations. IBM requires the department manager to notify the LSO prior to the procurement and construction of laser equipment.

The laser supervisor should notify the LSO of all personnel scheduled to work with potentially hazardous lasers, and provide the LSO with data for scheduling of required medical examinations.

Suspected or real accidents resulting from a laser under the supervisor's cognizance should be reported to the LSO by the supervisor as soon as possible. Further, he should assist as required in securing medical attention for any individual involved in a laser accident.

Employee

Employees working with lasers obviously bear a measure of responsibility in assuring a hazard-free operation. Key areas of employee responsibility follow:

- Authorization for laser usage
- Authorization for being in the presence of an operating laser
- Adherence to safety rules and procedures
- Reporting laser accidents

Employees should operate a laser only when so authorized by the laser supervisor.

The presence of an employee in the vicinity of a potentially hazardous operating laser should occur only with the laser supervisor's authorization.

Adherence to laser safety rules and procedures set forth under supervision, and hence by government and/or company authorities, is essential.

Occurrence of a suspected or real laser accident involving employees under the laser supervisor's cognizance shall be reported to the supervisor as soon as possible. If the supervisor cannot be readily notified, the employee should contact the LSO.

Documentation of Safety Procedures and Responsibilites

A final requirement in the organization of a satisfactory laser safety program is the necessity of a certain minimal documentation of safety procedures and responsibilities. With suitable written procedures and clear assignment of responsibilities, all persons involved in the laser operation can then have a clear understanding of the laser hazards and their responsibilities in controlling such hazards.

LASER SAFETY TRAINING PROGRAMS

A good laser safety training program can provide many of the elements needed to establish a well-managed laser safety program. As with the laser safety program itself, needs and requirements vary with the laser application and organization. Three laser safety training programs developed and used at IBM's Systems Product Division[1] at Poughkeepsie, N.Y. offer a broad range of flexibility suitable to most needs and requirements. Course content, major presentation features, advantages and disadvantages of each of the three IBM training programs are summarized in Table 11-1.

AUDIO/VISUAL SOURCES FOR TRAINING PROGRAMS

Obtaining suitable audio/visual aids for use in a laser safety training program can be a problem, as well as time-consuming. Several approaches and sources are suggested in the discussion that follows.

Professional organizations can be of much assistance. The Laser Institute of America (LIA) offers an 80-slide program with audio cassette narration, entitled "LIA Laser Safety Slide and Audio Cassette Educational Package Applications Slide Set." August 1977 price is $125 for LIA members and $250 for nonmembers.

Table 11-1. Summary of IBM Laser Safety Training Programs. [1]

Course Title	Contents of Course	Major Presentation Features	Advantages	Disadvantages

1. Comprehensive Expandable Classroom Program	*Session #1—Introduction*
	• Principle of operation of ruby laser.
	• Survey of other common lasers—wavelength, power/energy, operating modes.
	• Types of laser emission—continuous wave (CW), Q-switched, long pulse, etc.
	
	Session #2—Safety (Prime session of course)
	• Basics of biological effects on human organs caused by laser beam.
	• Safety eyewear.
	• Review of company and facility laser safety requirements.
	
	Session #3—Material Processing With Lasers
	• Laser applications in IBM and industry in general, with advantages of different lasers for particular applications.
	• Good and weak laser system design from a safety point of view.
		• Instructor (all four sessions).
		• Slides correlated with figures from two texts.
		• No mathematics or complicated relationships.
		
		• Irradiation of hot dog or chicken leg, for CO_2 laser operators.
		
		• 60–80 slides.
		• Hologram demonstration.
		• Display of laser processed parts (e.g, welded, drilled, cut, etc.).
			• Instructor control,
			  –allows discussion
			  –allows flexibility of course content and duration.
			• Reduced "On-the-Job" (O.J.T.) training phase with actual equipment.
				• Scheduling problems when instructors have other responsibilites.
				• Impractical to give course for limited number of students.
				• Differing instructor backgrounds results in variance of content and emphasis of laser safety session.

Table 11-1. (*Continued*)

Course Title	Contents of Course	Major Presentation Features	Advantages	Disadvantages
	Session #4– Systems Operations • Specifics on laser equipment student is to work with. • Presentation (via slides) of controls, adjustments, panels, etc. • Detailed methods of operation and operational procedures. • Description of cleaning and daily maintenance procedures. • Description of accessory equipment(s).	• Slides.		
2. Movie-Tape/ Slide Program	*Laser Fundamentals* *Laser Safety* • Basics of biological effects on eyes and skin caused by laser beam. • Laser beam and associated hazards, with emphasis on IBM's own laser uses and environment.	• Five minute excerpt of film "Fundamentals of Lasers" from Upjohn Vanguard series "Clinical Applications of Laser." • "Laser Safety" film from Extension Media Center of University of California (for development and laboratory lasers). • Slides	• Scheduling for any number of students on short notice. • Classroom instructor not required. • Standardization of material presented to student.	• No questions and answers without instructor. • No provisions for testing student. • No O.J.T. provisions.

	Control Measures			
3. Computer-Assisted Instruction (CAI)	• Engineering, personal protection and administrative controls, with emphasis on IBM's own laser uses and environment. • Same movies and slides used as Movie-Tape/Slide Program. • Special Section describing specific manager responsibilities. • Figure 11-1 shows the CAI course flow and includes objectives, teach and question sequence, review sequence and final quiz. To pass course all questions must be answered correctly; for missed questions a "remedial teach" sequence occurs.	• Some "Laser Safety" film as above. • Slides. • IBM 3270 display terminal with CRT display tube and keyboard. • Audio/visual unit with cassette recorder and 35mm slide projector (adjacent to computer terminal). • *Manager's Reference Manual.*	• No delay in training for small number of students. • Classroom instructor not required. • Much flexibility. Student can go at own pace with no peer pressure; take quiz immediately, or after review sequence, or go through complete teach and review sequence prior to quiz. • Uniform instructional material. • Uniform testing approach.	• No questions and answers without instructor. • No O.I.T. provisions.

[1] Summary compiled from information contained in Reference 1.

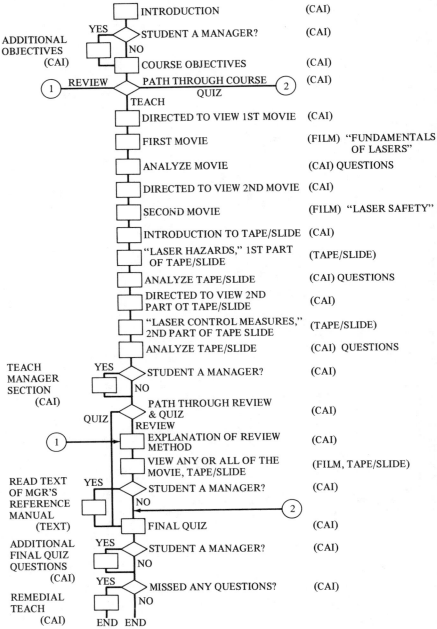

Figure 11-1. CAI laser safety course flow. *Note*: Figure 11-1 from Reference 1.

Film libraries provide another source of audio/visual training material. A goodly number of films on lasers and laser safety are available for rental, purchase or even borrowing. Several examples are cited in Table 11-1 under the "Major Presentation Features" column.

An often fruitful, although time-consuming, approach to obtaining slides is by contacting authors of technical publications and presentations.

REFERENCES

1. Smith, J. F., Murphy, J. J., Eberle, W. J., "Industrial Laser Safety Program Management," System Products Division, I.B.M. Corp., Poughkeepsie, N.Y.
2. Industrial Code Rule 50, "Lasers," State of New York, Department of Labor, Board of Standards and Appeals, Albany, N.Y., August 1, 1972.
3. "American National Standard for the Safe Use of Lasers, Z 136.1-1976", American National Standards Institute, New York, 1976.
4. "A Guide for Control of Laser Hazards," American Conference of Governmental Industrial Hygienists, Cincinnati, Ohio, 1973.

12

Safety in Classroom Laser Use

INTRODUCTION

The laser is commonly used as a demonstration device for teaching optics and wave mechanics. It has been shown to be of much value in high school or college courses in basic physics and optics.

The Southwestern Radiological Health Laboratory (SWRHL), a field laboratory of the Bureau of Radiological Health, U.S. Public Health Service, has been given responsibility for the technical implementation of Public Law 90-602, The Radiation Control for Health and Safety Act, with respect to lasers.[1] The SWRHL prepared a manual entitled *Laser Fundamentals and Experiments*,"[1] partially in response to the increased use of lasers for demonstration purposes in high schools and colleges, where potential exposure of large numbers of people is great. The manual is directed primarily toward the high school instructor who may use the laser in the classroom.

The material covered in this chapter deals with specific safety aids to help assure safe laser classroom usage, and has been adopted from Reference 1. It should be noted that demonstration lasers are addressed in the BRH's "Laser Products Performance Standards,"[2] and manufacturers are obliged to conform with its provisions as of August 1976. BRH clearly indicates that demonstration laser products are intended for demonstration, entertainment, advertising display or artistic composition.   Other types of manufactured lasers are not considered by the BRH to be demonstration lasers, even through the purchaser may choose to use it for such purposes. Further, demonstration lasers are Class I or Class II products, which are not eye-hazardous unless one stares into the beam of a Class II device. Thus, for recently manufactured demonstration lasers little if any hazard should exist for classroom usage. However, since a large number of potentially hazardous lasers may still be in classroom use, and since other than demonstration-type lasers could quite conceivably find their way into the classroom, the material covered in this chapter remains of much relevance.

SAFETY AIDS

Laser light can pose a definite hazard to the eye, but can also be used safely in the classroom. Common sense and preplanning of experiments will serve to point out most obvious hazards. Safety aids will help prevent injury even when the unexpected does occur. The following represent useful safety aids for classroom laser usage:

- The beam shutter
- The target
- The demonstration box
- The display tank
- Black paint
- Reduction of beam intensity
- Key lock
- The dry run

The Beam Shutter

Lasers are constructed to withstand continuous use, and their life span is actually shortened by intermittent usage. Turning the laser on and off at short intervals is also inconvenient. One way to overcome this inconvenience is by using a beam shutter.

The shutter can be a simple mechanical device fitted over the aperture of the laser, allowing the operator to cut off the laser beam without actually turning off the laser. The shutter should be made of

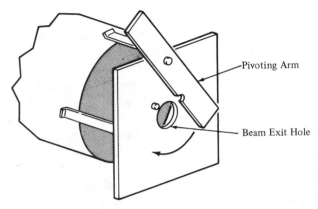

Figure 12-1. Beam shutter. Source: BRH, Southwestern Radiological Health Laboratory, Reference 1.

black, nonreflective material, and should completely stop passage of the laser beam. An example of a simple beam shutter device is shown in Figure 12-1.

Use of the beam shutter can reduce the hazard to the operator whenever the experimental configuration in front of the laser is being changed or altered. The beam shutter eliminates the possiblity of accidental reflections from a piece of equipment during the changing process. As with all safety aids, one must develop the habit of shuttering the beam at all times that it is not actually needed.

The Target

The laser beam travels outward from the laser until absorbed or reflected. To prevent accidents caused by reflections from specular surfaces, a target of suitable material should be provided for the beam. The target should be made of nonreflective material and be large enough to stop the beam under a wide variety of experimental situations. Black foam rubber material and black ink on a blotter are good examples of targets.

The Demonstration Box

Accidents can occur during setup and alignment of the laser demonstration. Observers should be protected during this phase of the demonstration by placing a shield between the laser and the observers.

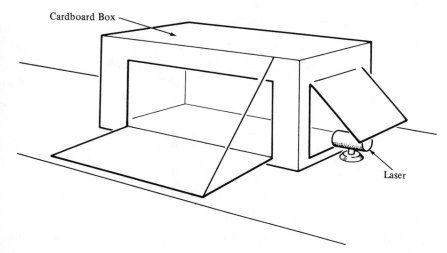

Figure 12-2. Demonstration box. Source: BRH, Southwestern Radiological Health Laboratory, Reference 1.

One approach is to hang a black, dull-surface curtain or drape between the observers and the demonstration.

Another simple and inexpensive technique is to use a large cardboard box with holes in either end for entrance and, if necessary, exit of the laser beam. Large panels can be cut out of either side of the box, one for use while setting up the experiment, another for viewing. Doors or flaps can cover the opening when it is not in use. The demonstration box described is shown in Figure 12-2.

The Display Tank

Laser light is invisible from the side unless it is scattered by a medium. A convenient and relatively safe way of viewing the beam is to utilize a clear, liquid-filled display tank. The tank can be easily constructed of Plexiglass or any clear, thick plastic. Surfaces should be flat and square with one another to permit accurate passage of the beam through the tank. The liquid filling can be comprised of a solution of a clear substance with some large molecules. Transmission oil may be acceptable, or a soap solution with some red food coloring mixed in. Caution must be exercised in setting up the display tank, as reflections from the tank sides could be hazardous. See Figure 12-3 for a display tank drawing.

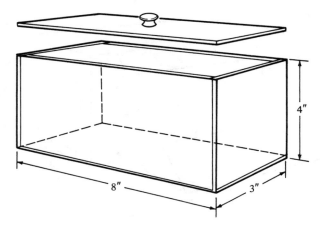

Figure 12-3. Display tank. Source: BRH, Southwestern Radiological Health Laboratory, Reference 1.

Use of a jar or a bottle as a display tank is not advisable. Inexact alignment of the beam could result in the beam's being specularly reflected from the side of the bottle.

A second, and perhaps more convenient, display device can be fabricated from plastic casting resin such as is found in hobby stores. When cast into a square or rectangular block, with one face painted flat black, the resin will serve well to display a beam passing through it.

Black Paint

Painting the surfaces of demonstration equipment with a flat, dull black paint may be the most important single precaution one can take, since it prevents specular reflections. It is unfortunate from a laser safety point of view that many pieces of optical gear are supplied with bright plated surfaces. Stainless steel and chrome are quite attractive, but the specular reflections from these surfaces can be literally blinding.

Reduction of Beam Intensity

Many lasers available for educational use furnish far more light than is needed to conduct a demonstration adequately. The power density

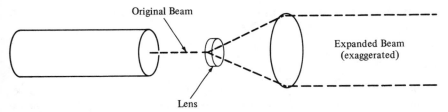

Figure 12-4. Beam expander. Source: BRH, Southwestern Radiological Health Laboratory, Reference 1.

of the beam should be reduced to a level commensurate with the level of laser light actually required. This precaution reduces the potential for accidental eye damage. Two methods work well for reducing the beam intensity.

The first method is simply to insert an absorbing filter in the path of the beam. This filter may be a neutral density type. Any other material that effectively absorbs the light while at the same time it does not scatter the beam, will also function satisfactorily.

The second method is to increase the diameter of the beam by using a pair of lenses. The first lens serves to spread the beam. The second lens functions to recollimate the beam to near its original divergence. Expanding the beam reduces the power density and lessens the hazard posed by direct or indirect viewing of the beam. If feasible, permanent attachment of the lenses to the laser is recommended. It should be noted that expanding the beam will not destroy the coherent properties of laser light. A simplified drawing of a beam expander is shown in Figure 12-4.

Key Lock

If possible, the classroom laser should be purchased with a key lock for the power supply. If that is not possible, then installation of a key lock should be implemented. With a key-lock arrangement, unauthorized use of the laser can be readily prevented by the instructor.

The Dry Run

All laser experiments should be dry-run before presenting them to the class. This permits the operator to determine where the hazards are, or may be, and allow him to eliminate them before they can cause damage.

REFERENCES

1. VanPelt, W. F., Stewart, H. F., Peterson, R. W., Roberts, A. M., Worst, J. K., *Laser Fundamentals and Experiments*, U.S. Dept. of Health, Education and Welfare, Public Health Service, Bureau of Radiological Health, Southwestern Radiological Health Laboratory, Las Vegas, Nevada, May 1970.
2. "Laser Products Performance Standards," *Federal Register*, Vol. 40 No. 148, pp. 32252–32266, U.S. Dept. of Health, Education and Welfare, Food and Drug Administration, Bureau of Radiological Health (BRH), Washington, D.C., 31 July 1975.

13
Medical Surveillance

INTRODUCTION

As has been discussed in Chapter 11, Laser Safety Program, individuals working with lasers require some measure of medical surveillance as part of a sound laser safety program. This chapter addresses the details of such a medical surveillance program.

Except for very high powered lasers, where internal organs can be destroyed or seriously damaged, the two organs normally susceptible to laser radiation damage are the eyes and the skin. Thus, medical surveillance includes ophthalmological and dermatological considerations. It follows that qualified medical personnel should perform the required examinations for implementing a suitable medical surveillance program.

There are two prime purposes for medical surveillance of personnel working with lasers:

- To formulate a baseline on the condition of the eyes and skin before working with lasers.
- To detect and document at the earliest time possible, any ocular or skin damage that may occur as a result of exposure to laser radiation.

A number of benefits can accrue. Overall, the effectiveness of the control measures can be judged, and they can be modified if necessary. Long-term exposure effects might become evident with continued examinations and by keeping good records. Suitable therapeutic measures can be quickly taken, should that be necessary. Finally, the possibility of attributing damage (in particular to the eye) to laser radiation when a pathological condition might have existed prior to working with lasers can be ruled out with a suitable baseline examination.

Medical surveillance is not needed for persons working with Class I, Class II or Class IIIa lasers.

These relevant topics are covered in the sections that follow:

- Personnel risk classification
- Types of eye examinations
- Skin surveillance requirements
- Recommended medical examination requirements
- Frequency of medical examinations

PERSONNEL RISK CLASSIFICATION

The American National Standards Institute[1] categorizes personnel who should be under laser medical surveillance for possible eye or skin damage into three risk classifications:

- Minimal risk personnel
- Moderate risk personnel
- High risk personnel

Minimal risk personnel are defined as those whose work makes it possible but unlikely that they would be exposed to laser energy or power sufficient to damage their eyes or skin.

Moderate risk personnel regularly work in laser environments, but are usually fully protected by procedural control measures or controls built into the equipment.

High risk personnel run a high risk of being exposed to sufficiently intense laser radiation to damage the eyes and/or skin.

As noted in Chapter 11, Laser Safety Program, the Laser Safety Officer (LSO) is responsible for having the necessary medical examinations carried out and records kept. ANSI[1] recommends that the risk classification be determined by the LSO in charge of the laser installation.

TYPES OF EYE EXAMINATIONS

The extent of the eye examination is governed by the personnel risk classification of those working with lasers. Table 13-1 describes a number of types of eye examinations that may be required. Ocular history is included as an "eye examination" item.

Table 13-1. Summary of Types of Eye Examinations.

Item No.	Title of Examination	Description of Examination
1	Ocular history	• Review patient's past eye history and family eye history. • Note any current complaints patient has about eyes. • Inquire about patient's general health status, with special emphasis on diseases that can given ocular problems. • Record patient's present lens prescription, if any.
2	Visual acuity	• Test and record visual acuity in Snellen figures for 20 feet with and without lenses to 20/15. • Visual acuity at near is tested at 35 cm and recorded in Jaeger-test figures, with and without lenses.
3	External ocular examination	Includes examination of • Brows • Lids • Lashes • Conjuctiva • Sclera • Cornea • Iris • Pupillary size • Equality reactivity and regularity
4	Amsler grid	• The Amsler grid sheet is presented to each eye separately. • Any distortion of the grid is noted by the patient and drawn by him.
5	Manifest refraction	• This exam is to measure the patient's refractive error. The new visual acuity of the patient must be noted if the visual acuity is improved over that achieved with the patient's old lens prescription. Also, note if he has no lenses at the time of the examination. • Examination to be carried out in all personnel whose visual acuity in either eye is less than 20/20.

Table 13-1. (*Continued*)

Item No.	Title of Examination	Description of Examination
6	Measurement of intraocular pressure	• Pupils are then dilated by the instillation of a mydriatic drop in each eye. • Remainder of examination is carried out with the eye under this medication.
7	Examination by slit lamp	• The cornea, iris and lens are examined with this biomicroscope (e.g., slit lamp) and described.
8	Examination of the ocular fundus with an ophthalmoscope	• In the recording of this portion of the examination the points to be covered follow: —Presence or absence of opacities in the media. —Sharpness of outline of the optic nerve. —Size of the physiological cup if present. —Ratio of the size of the retinal veins to that of the retinal arteries. —Presence or absence of a well-defined macula and the presence or absence of a foveolar reflex. —Any retinal pathology that can be seen with a direct ophthalmoscope. • Even small deviations from normal should be described and carefully localized.
9	Photograph of the posterior pole of the fundus	• Includes the area of the macula and head of the optic nerve, and is to be taken in color film. • A positive color photograph should be mounted in the patient's chart.
10	Other examinations	Further examinations may be done as deemed necessary by the examiner.

Note: Table 13-1 compiled from Reference 1.

SKIN SURVEILLANCE REQUIREMENTS

Medical surveillance of the skin is required only of high risk classification personnel. The skin surveillance includes the following three areas:

- History
- Examination
- Recommendations

Table 13-2 summarizes the content of the history and examination portions of the skin surveillance.

Table 13-2. Summary of Skin Surveillance History and Examination Requirements.

Item No.	General Requirements	Description of Requirement
1	History	• Atropy • History of exposure of X-ray, nuclear radiation, ultraviolet sources, high intensity visible or infrared sources, arsenic, cutting oils and antibiotics. • Absence or presence of reactions to drugs and chemicals. • Photodermatitis. • Intensity of exposure to sun. • Skin cancers, including type, location and treatment. • History of previous skin disorders. • History of extensive or severe thermal burns.
2	Examination A) Skin Color	A) Skin color is to be noted as to: • Caucasian—light, medium or dark • Negro—light, medium or dark • Oriental—light, medium or dark Should special instrumentation be available for precise analysis of skin color, it should be used and the data recorded.
	B) Presence or Absence of Skin Changes	B) Presence or absence of the following changes should be observed and noted under normal light and Wood's light on exposed and unexposed skin (body charts may be used): • General skin type: normal (N), seborrheic (S) or ichthyotic (I) • Pigmentation: type, severity, distribution • Depigmentation: type, severity, distribution • Keratoses: actinic, seborrheic; type, distribution patterns • Skin malignancies: type, size, distribution patterns • Skin disorders: type, distribution patterns
	C) Photographs	C) Photographs should be made if indicated, including black-and-white, color, and under Wood's light.

Note: Table 13-2 compiled from Reference 1.

Specific recommendations should be made regarding at least the following areas:

• Skin protection needs, for personnel control and/or area control.
• Use of gloves
• Protective creams.

- No laser exposure, limited laser exposure or laser exposure with ordinary precautions.
- Need for annual skin examination.
- Need for final examination for high risk personnel upon leaving laser work.

RECOMMENDED MEDICAL EXAMINATION REQUIREMENTS

Medical examination requirements are based on ANSI[1] recommendations, and are correlated with the personnel risk classifications discussed earlier.

Minimal Risk Personnel Examination Requirements

Personnel of this type require only an eye examination for visual acuity. See Item 2 of Table 13-1 for details of the visual acuity examination.

No skin medical surveillance is necessary for minimal risk personnel.

Moderate Risk Personnel Examination Requirements

This group requires a minimal eye examination covering Items 1, 2 and 8 of Table 13-1, namely:

- Ocular history.
- Visual acuity.
- Examination of the ocular fundus with an ophthalmoscope, and pupils dilated.

If visual acuity in either eye is less than 20/20 with corrective lenses, or if any pathology is seen, it is recommended that the prospective laser worker have a complete eye examination, as is required for high risk personnel.

No skin medical surveillance is necessary for moderate risk personnel.

High Risk Personnel Examination Requirements

Personnel of this type require a complete eye examination covering all items of Table 13-1.

A complete skin examination and skin history are required as summarized in Table 13-2.

FREQUENCY OF MEDICAL EXAMINATIONS

The frequency of medical examinations recommended by ANSI[1] is based on the personnel risk classification.

Minimal risk personnel should have the visual acuity examination done prior to starting work with lasers.

Moderate risk personnel should undergo the needed medical examinations at the following times:

- Prior to starting laser work.
- After completion of laser work.
- Right after suspected or actual eye or skin injury due to laser radiation exposure.

For high risk personnel, the required medical examinations should be done at the following times:

- Prior to starting laser work.
- After completion of laser work.
- Right after suspected or actual eye or skin injury due to laser radiation exposure.
- At three-year intervals while working with lasers.

REFERENCES

1. "American National Standard for the Safe Use of Lasers, Z 136.1-1976," American National Standards Institute, New York, 1976.

BIBLIOGRAPHY

1. "Control of Hazards to Health from Laser Radiation," U.S. Department of the Army Technical Bulletin TB-MED-279, Washington, D.C., May 1975.
2. Goldman, L., Rockwell, R. J., Jr., Hornby, P., "Laser Laboratory Design and Personnel Protection from High Energy Lasers," CRC Handbook of Laboratory Safety, N. V. Steere, Editor, 2nd Edition, Chemical Rubber Co., Cleveland, Ohio, pp. 381–389, 1971.

14

Laser Protective Eyewear

INTRODUCTION

The need for laser protective eyewear arises when a sufficiently high risk exists that exposure to potentially damaging laser radiation is possible, after all reasonable control measures have been taken. This is generally the case for laboratory laser systems, where ready access to system components is required by the personnel conducting the experiments.

This chapter addresses the following aspects of laser protective eyewear:

- Factors in selecting protective eyewear
- Identification of eyewear
- Inspection of eyewear
- Types of protective eyewear
- Responsibility of manufacturer
- Broadband development

FACTORS IN SELECTING PROTECTIVE EYEWEAR

Many factors come into play in selecting protective eyewear. These factors include:

- Wavelength of laser output
- Laser beam intensity
- Maximum Permissible Exposure (MPE)
- Optical Density (D_λ or O.D.) of eyewear
- Visible (or luminous) transmittance
- Laser filter radiation damage threshold
- Comfort and fit
- Operational requirements for laser eye protection
- Other physical characteristics
- Requirement for corrective glasses
- Requirement for peripheral vision

Wavelength of Laser Output

The laser radiation wavelength(s) is a key factor governing the type of protective eyewear to be used to prevent such radiation from reaching the eye.

A number of lasers emit more than a single wavelength. Each wavelength or spectral range must be considered on its own. Concern only with the more intense radiation can be dangerous. Protective eyewear must attenuate each hazardous spectral output or additive outputs to eye-safe levels.

Laser Beam Intensity

A second key laser output parameter that must be known for appropriate selection of safety eyewear is the laser beam intensity. For pulsed lasers, it is the maximum radiant exposure in $J \cdot cm^{-2}$ that must be known, and maximum irradiance in $W \cdot cm^{-2}$ for CW lasers.

Maximum Permissible Exposure (MPE)

For a given laser beam intensity, wavelength and exposure duration, selection of suitable protective eyewear requires a knowledge of the respective MPE. (See Chapter 6, Protective Standards, for MPE exposure criteria for the eye.) With this information it then becomes possible to assign an attenuation value that the eyewear must provide for safe usage.

Optical Density (D_λ or O.D.) of Eyewear

Optical density is an attenuation parameter. For a particular exposure (H), the optical density that the protective eyewear must provide in order to attenuate the level to an eye-safe MPE value is given by,

$$\text{O.D.} = \log_{10} \frac{H}{MPE} \qquad \text{(Eq. 14-1)}$$

For H, use the same units as those of the appropriate MPE. (See Chapter 6 for MPE exposure criteria for the eye.)

Table 14-1 lists several values of the minimum optical density that protective eyewear would have to provide for particular values of H/MPE.

Laser beam intensities may be many thousands of times above safe exposure levels (MPEs). Thus, dealing with logarithmic notation of optical density is far less awkward than having to work with percent transmission notation. For example, protective eyewear with a transmission of 0.0001% is much more pleasingly described as having an optical density of 6. Table 14-2 provides corresponding values of optical density vs. percent transmission.

Table 14-3 provides a simplified approach for determining the optical density requirements for selected exposure conditions at wavelengths between 200 and 1,400 nm.

The optical density vs. wavelength curves of Figures 14-1 to 14-4 are representative of the performance of present-day laser protective eyewear. It can be seen from these curves that protection is

Table 14-1. Minimum Optical Densities Required of Protective Eyewear.

$\dfrac{H}{MPE}$	O.D.
$1 = 10^0$	0
$10 = 10^1$	1
$100 = 10^2$	2
$1,000 = 10^3$	3
$10,000 = 10^4$	4
$100,000 = 10^5$	5
$1,000,000 = 10^6$	6
$10,000,000 = 10^7$	7
$100,000,000 = 10^8$	8

Table 14-2. Percent Transmission vs. Optical Density.

Percent	Density	Percent	Density	Percent	Density	Percent	Density	Percent	Density
100.0	0.00	33.11	0.48	10.96	0.96	3.631	1.44	1.20	1.92
97.72	0.01	32.36	0.49	10.72	0.97	3.548	1.45	1.17	1.93
95.50	0.02	31.62	0.50	10.47	0.98	3.467	1.46	1.15	1.94
93.33	0.03	30.90	0.51	10.23	0.99	3.388	1.47	1.12	1.95
91.20	0.04	30.20	0.52	10.00	1.00	3.311	1.48	1.10	1.96
89.13	0.05	29.51	0.53	9.772	1.01	3.236	1.49	1.07	1.97
87.10	0.06	28.84	0.54	9.550	1.02	3.162	1.50	1.05	1.98
85.11	0.07	28.18	0.55	9.333	1.03	3.090	1.51	1.02	1.99
83.18	0.08	27.54	0.56	9.120	1.04	3.020	1.52	1.00	2.00
81.28	0.09	26.92	0.57	8.913	1.05	2.951	1.53	0.89	2.05
79.43	0.10	26.30	0.58	8.710	1.06	2.884	1.54	0.79	2.10
77.62	0.11	25.70	0.59	8.511	1.07	2.818	1.55	0.71	2.15
75.85	0.12	25.12	0.60	8.318	1.08	2.754	1.56	0.63	2.20
74.13	0.13	24.55	0.61	8.128	1.09	2.692	1.57	0.56	2.25
72.44	0.14	23.99	0.62	7.943	1.10	2.630	1.58	0.50	2.30
70.79	0.15	23.44	0.63	7.762	1.11	2.570	1.59	0.45	2.35
69.18	0.16	22.91	0.64	7.586	1.12	2.512	1.60	0.40	2.40
67.61	0.17	22.39	0.65	7.413	1.13	2.455	1.61	0.36	2.45
66.07	0.18	21.88	0.66	7.244	1.14	2.399	1.62	0.32	2.50
64.57	0.19	21.38	0.67	7.079	1.15	2.344	1.63	0.28	2.55
63.10	0.20	20.89	0.68	6.918	1.16	2.291	1.64	0.25	2.60
61.66	0.21	20.42	0.69	6.761	1.17	2.239	1.65	0.22	2.65
60.26	0.22	19.95	0.70	6.607	1.18	2.188	1.66	0.20	2.70
58.88	0.23	19.50	0.71	6.457	1.19	2.138	1.67	0.18	2.75
57.54	0.24	19.05	0.72	6.310	1.20	2.09	1.68	0.16	2.80
56.23	0.25	18.62	0.73	6.166	1.21	2.04	1.69	0.14	2.85
54.95	0.26	18.20	0.74	6.026	1.22	2.00	1.70	0.13	2.90
53.70	0.27	17.78	0.75	5.888	1.23	1.95	1.71	0.11	2.95
52.48	0.28	17.38	0.76	5.754	1.24	1.91	1.72	0.10	3.00
51.29	0.29	16.98	0.77	5.623	1.25	1.86	1.73	0.09	3.04
50.12	0.30	16.60	0.78	5.495	1.26	1.82	1.74	0.08	3.10
48.98	0.31	16.22	0.79	5.370	1.27	1.78	1.75	0.07	3.15
47.86	0.32	15.85	0.80	5.248	1.28	1.74	1.76	0.06	3.20
46.77	0.33	15.49	0.81	5.129	1.29	1.70	1.77	0.04	3.40
45.71	0.34	15.14	0.82	5.012	1.30	1.66	1.78	0.025	3.60
44.67	0.35	14.79	0.83	4.898	1.31	1.62	1.79	0.016	3.80
43.65	0.36	14.45	0.84	4.786	1.32	1.58	1.80	0.010	4.00
42.66	0.37	14.13	0.85	4.677	1.33	1.55	1.81	0.006	4.25
41.69	0.38	13.80	0.86	4.571	1.34	1.51	1.82	0.003	4.50
40.74	0.39	13.49	0.87	4.467	1.35	1.48	1.83	0.0018	4.75
39.81	0.40	13.18	0.88	4.365	1.36	1.45	1.84	0.0010	5.00
38.90	0.41	12.88	0.89	4.266	1.37	1.42	1.85	0.0006	5.25
38.02	0.42	12.59	0.90	4.169	1.38	1.38	1.86	0.0003	5.50
37.15	0.43	12.30	0.91	4.074	1.39	1.35	1.87	0.00018	5.75
36.31	0.44	12.02	0.92	3.981	1.40	1.32	1.88	0.00010	6.00
35.48	0.45	11.75	0.93	3.890	1.41	1.29	1.89		
34.67	0.46	11.48	0.94	3.802	1.42	1.26	1.90		
33.88	0.47	11.22	0.95	3.715	1.43	1.23	1.91		

Source: U.S. Army Environmental Hygiene Agency, Reference 1.

Table 14-3. Simplified Method for Selecting Laser Eye Protection for Intrabeam Viewing for Wavelengths Between 200 and 1,400 nm.

Q-Switched Lasers (1 ns to 0.1 ms)		Non-Q-Switched Lasers (0.4 ms to 10 ms)		Continuous Lasers Momentary (0.25 s to 10 s)		Continuous Lasers Long-Term Staring Greater than 3 hr		Attenuation	
Maximum Output Energy (J)	Maximum Beam Radiant Exposure (J·cm⁻²)	Maximum Laser Output Energy (J)	Maximum Beam Radiant Exposure (J·cm⁻²)	Maximum Power Output (W)	Maximum Beam Irradiance (W·cm⁻²)	Maximum Power Output (W)	Maximum Beam Irradiance (W·cm⁻²)	Attenuation Factor	O.D.
10	20	100	200	NR	NR	100	200	100,000,000	8
1.0	2	10	20	NR	NR	10	20	10,000,000	7
10^{-1}	2×10^{-1}	1.0	2	10^{3}	2×10^{3}	1.0	2	1,000,000	6
10^{-2}	2×10^{-2}	10^{-1}	2×10^{-1}	100	200	10^{-1}	2×10^{-1}	100,000	5
10^{-3}	2×10^{-3}	10^{-2}	2×10^{-2}	10	20	10^{-2}	2×10^{-2}	10,000	4
10^{-4}	2×10^{-4}	10^{-3}	2×10^{-3}	1.0	2	10^{-3}	2×10^{-3}	1,000	3
10^{-5}	2×10^{-5}	10^{-4}	2×10^{-4}	10^{-1}	2×10^{-1}	10^{-4}	2×10^{-4}	100	2
10^{-6}	2×10^{-6}	10^{-5}	2×10^{-5}	10^{-2}	2×10^{-2}	10^{-5}	2×10^{-5}	10	1

Source: U.S. Army Environmental Hygiene Agency, Reference 1.

Note: NR = not recommended.

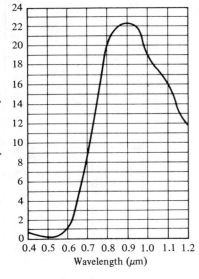

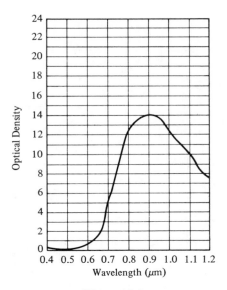

Figure 14-1.   **Figure 14-2.**

Figure 14-1. Optical density curve of filter used primarily with the ruby laser and other lasers in the near infrared.

Notes:
(1) Figure 14-1 adapted from Reference 2.
(2) Luminous transmittance is 35%.
(3)

Laser	Wavelength	O.D.
HeNe	0.633 μm	2.7
Ruby	0.694 μm	8.0
GaAs	0.84 μm	20.0
Nd	1.06 μm	17.1

Figure 14-2. Optical density curve of filter used primarily with the neodymium doped laser and other lasers in the near infrared.

Notes:
(1) Figure 14-2 adapted from Reference 2.
(2) Luminous transmittance is 46%.
(3)

Laser	Wavelength	O.D.
GaAs	0.84 μm	14.6
Nd	1.06 μm	11.9

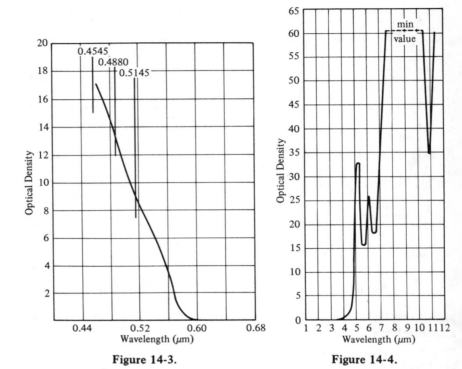

Figure 14-3.   **Figure 14-4.**

Figure 14-3. Optical density curve of filter used primarily with the argon ion laser and other lasers in the ultraviolet.

Notes:

(1) Figure 14-3 adapted from Reference 2.

(2) Luminous transmittance is 24%.

(3)
Laser	Wavelength	O.D.
Argon	0.4545 μm	17.7
ion	0.488 μm	13.5
	0.5145 μm	9.0

Figure 14-4. Optical density curve of filter used primarily with the carbon dioxide laser and other lasers in the infrared.

Notes:

(1) Figure 14-4 adapted from Reference 2.

(2) Luminous transmittance is almost 100%.

(3) Optical density of CO_2 laser wavelength (10.6 μm) is 50.

available for today's most widely used lasers. Curves of O.D. vs. wavelength for Glendale Optical Company's LGS and LGU series goggles are shown in Figures 14-5 through 14-9.

Transmission data for many filter materials typically exhibit sharp drops from high attenuation to complete transmission. Thus, extrapolation of optical density/wavelength values should not be attempted without supporting test data, when it comes to designing or selecting special protective eyewear.

For higher-intensity lasers, care must be exercised to assure that skin MPEs are not exceeded, even though protective eyewear optical density satisfactorily reduces radiation levels below eye MPE values.

The optical density of two absorbing filters when placed side by side is approximately the sum of the two individual optical densities.

Visible (or Luminous) Transmittance

The ability of an individual to see through an optical filter (e.g., protective lens) can be described in terms of luminous transmittance, and is usually expressed in terms of percent.

While adequate optical density must be the overriding consideration in the selection of laser protective eyewear, the need for sufficient visible transmittance must be weighed. A low visible transmittance can create problems of eye fatigue, and can be particularly hazardous in a laboratory environment.

Sunglasses have luminous transmittance values typically between 10 and 40%, as do most laser eye-protective filters.[2] However, sunglasses do not have adequate attenuation at laser wavelengths to provide useful protection.

Laser Filter Radiation Damage Threshold

Exposure of laser eyewear filter materials to a sufficiently intense laser beam will damage the filter, and damage can take the form of bleaching, bubbling, melting or shattering. R. L. Elder[3] of the Bureau of Radiological Health (BRH) indicates that tests have shown some eye safety materials to fail after several seconds of exposure to laser beams of approximately 1 watt, or power density of about

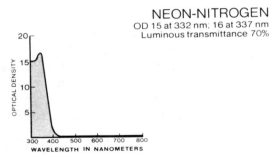

Figure 14-5. Neon-nitrogen curve. (Courtesy of Glendale Optical Co.)

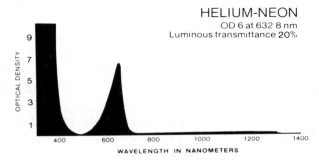

Figure 14-6. Helium-neon curve. (Courtesy of Glendale Optical Co.)

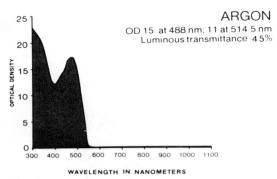

Figure 14-7. Argon curve. (Courtesy of Glendale Optical Co.)

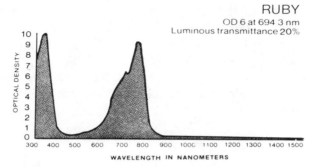

Figure 14-8. Ruby curve. (Courtesy of Glendale Optical Co.)

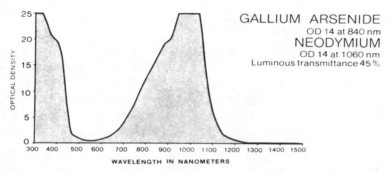

Figure 14-9. Gallium arsenide and neodymium curve. (Courtesy of Glendale Optical Co.)

6–12 W · cm^{-2}. The U.S. Army Environmental Hygiene Agency[1] indicates typical damage thresholds from Q-switched pulsed laser radiation to be between 10 and 100 J · cm^{-2} for absorbing glass and 1 and 100 J · cm^{-2} for plastics and dielectric coatings. They further note that irradiances from CW lasers which would cause filter damage exceed levels that would present a serious fire hazard.

Personnel using protective eyewear in the proximity of high power lasers (e.g., a multiwatt argon-ion laser) should contact the manufacturer for determination of the failure points of the eyewear.[3] Should these data not be available, the eyewear should be tested by subjecting it to the most intense radiation expected to be encountered. The test would then be indicative of the worst-case condition. A careful inspection of the eyewear must be made before returning it to use. Further, should the eyewear fail, then a halt in lasing operations is required until an alternate approach to personnel safety is devised. Finally, protective eyewear should always be inspected before usage for indications of bleaching, bubbling, melting or cracking. Any of these signs should be cause for its immediate removal from use.

Comfort and Fit

Comfort and fit constitute another important factor requiring consideration when selecting laser safety eyewear. The protective eyewear must provide a snug fit. Otherwise, unattenuated laser radiation may reach the person's eyes.

A trade-off often must be made between selecting spectacle-type frames, which generally give more comfort, and a cover-all type of goggle, which gives a close fit. (See the next section on types of protective eyewear, and Figures 14-10 to 14-13.) While the cover-all frame is usually heavier and has a more restricted field of view than spectacle frames, it does offer the following advantages:

- It fits more closely than spectacles.
- It can be worn over spectacles.
- It can accomodate more complex filter arrangements.

However, the increased comfort of spectacle frames can in practice mean the difference between wearing the safety eyewear and not wearing it.

Figure 14-10. Plano spectacles (Courtesy of Glendale Optical Co.).

Figure 14-11. Spectacle clip-on (Courtesy of Glendale Optical Co.).

Figure 14-12. Goggles (Courtesy of Glendale Optical Co.).

Figure 14-13. Goggles (Courtesy of Glendale Optical Co.).

Operational Requirements for Laser Eye Protection

Requirements for laser eye protection can vary considerably, and necessitates good judgment and experience. The U.S. Army Environmental Hygiene Agency,[1] which appears to be the clearing house for Army laser safety considerations, addresses the general requirements or approaches for laser eye protection usage for the following operational situations, when potentially hazardous lasers are used:

- For flight crews
- For armored vehicle crews
- In test and training activities
- For indoor shop or laboratory environments

Eye protection for flight crews of aircraft furnished with laser rangefinders and target designators is normally not recommended. The fundamental rationale is that a very low probability for exposure from a reflected laser beam exists, and the disadvantages and hazards listed below may outweigh the protection obtained when wearing protective eyewear:

- Loss of peripheral vision
- Reduced visual transmission
- Degraded color contrast from most types of goggles

If, however, hazardous specular reflections are likely to be reflected towards the aircraft, then flight crews can be required to wear high-visibility protective eyewear.

Crews of armored vehicles do not require protective eyewear inside the vehicle unless devices such as optical sights or view blocks might transmit the beam to a crew member. Such devices should be fitted with filters for attenuating the laser beam. For crew members outside the vehicle, eye protection might be required if hazardous specular reflections are anticipated. Where the armored vehicle is the target in laser exercises, then protective eyewear is mandatory for the driver, the commander and other exposed personnel.

For test and training activities, protective eyewear must be worn by downrange individuals within the laser beam target area. It is also required for other personnel if specular surfaces cannot be removed from the target area. Removal of specular objects is more desirable, as it eliminates the need for eye protection for all but target-area personnel.

For indoor shop or laboratory usage requiring high-intensity lasers, protective eyewear is required if the beam is not enclosed. Eye protection is not needed for holographic viewing and optical alignment procedures, provided reasonable precautions are taken.

Other Physical Characteristics

In addition to its offering a high optical density and high luminous transmittance, other physical characteristics of the laser protective eyewear must be considered. These include:

- Mechanical durability
- Scratch and abrasion resistance
- General optical quality
- Stability of properties upon exposure to:
  −Light
  −Heat
  −Humidity

A special trait that laser protective eyewear must exhibit is that of resistance to laser irradiation. This characteristic can be attained by manufacturers through one or more of the following approaches:

- Choice of material
- Use of two or more filters in the protective goggle
- Placing a reflective coating over the outermost surface of the filter

Requirement for Corrective Glasses

Personnel requiring corrective glasses who work with lasers on a regular basis where eye protection is necessary (e.g., laser laboratory), should when possible secure laser protective and corrective eyewear. This type is generally most comfortable and discourages working in the environment in an unprotected fashion.

Requirement for Peripheral Vision

For flight crews, particularly pilots, loss of peripheral vision may cause undue hazard to the exercise or mission. Thus, the hazard must be weighed against the protection afforded by safety eyewear and the low probability that exists for exposure from reflected laser beams.

IDENTIFICATION OF EYEWEAR

It is essential that all laser protective eyewear be clearly marked to ensure that it is used for protection against the laser wavelength for which it is intended. Conversely, clear marking helps to assure that it is not used against laser wavelengths for which it is not intended. Thus, labeling of protective eyewear should clearly indicate the optical density value(s) and wavelength value(s) for which protection is provided. Such labeling is indeed a firm requirement of New York State Industrial Code Rule 50 for lasers.[4]

INSPECTION OF EYEWEAR

Visual inspections should be performed prior to each lasing to assure the basic integrity of the protective eyewear. At least the following checks should be made:

- Inspection of the attenuator filter material for bleaching, cracking, pitting, bubbling or melting.
- Inspection of the goggle frame for mechanical integrity and light leaks.
- Inspection of filter material for scratches.

New York State Industrial Code Rule 50 for lasers[4] specifies the following inspection requirements:

- Eyewear must be inspected at least every six months to assure that optical density is appropriate to the laser in use, and to determine visually that no optical defects exist.
- Safety eyewear that has been exposed to high-intensity laser radiation shall not be reused until it has been reevaluated for shattering and effectiveness of its attenuation.

TYPES OF PROTECTIVE EYEWEAR

Three basic types of laser protective eyewear exist:

- Corrective (or plano) spectacles. (See Figure 14-10.)
- One-piece visitor's spectacles, which can be used over corrective spectacles, are of light-weight construction and have side-shield protection. (See Figure 14-11.)
- Goggles, which have a covered plastic eyeshield and are generally the least comfortable eyewear. (See Figures 14-12 and 14-13.)

Table 14-4 summarizes Winburn's[5] recommendations for protective eyewear adopted by the Laser Research and Technology Division of the Los Alamos Scientific Laboratory. The only wavelength range not covered is 0.55–0.60 µm, a region of retinal sensitivity. Filters exist, but visual transmission is an extremely low 5%.[5] Thus, in that case the risk of other laboratory hazards (e.g., electrical shock) associated with laser operation is enhanced. Complete enclosure of 0.55–0.60-µm lasers must therefore be strongly advised.

RESPONSIBILITY OF MANUFACTURER

ANSI[6] recommends that manufacturers of laser protective eyewear provide the following information with each item:

- Wavelength(s) and corresponding optical density for which protection is provided.
- Any other pertinent data for laser safety purposes.

The U.S. Army Environmental Hygiene Agency[1] notes that, in general, goggles meet or exceed the specification given by the manufacturer. However, on rare occasions protective filters have been determined to have optical densities less than specified. It is further noted by the Agency that all evidence indicates that optical density of commercially available eyewear does not decrease with age.

BROADBAND DEVELOPMENT

A difficult safety problem arises with tunable lasers that operate over a broad range of wavelengths.

Table 14-4. Properties of Recommended Laser Protective Eyewear.[1,5]

Wavelength Range (μm)	Specific Type	Laser Wavelength (μm)	Manufacturer	Model Designation	Type	Optical Density	Visible Transmission (%)	Cost
0.22–0.42	—	UV	Ultra-Violet Products, Inc.	UVC-303	One piece visitor's spectacles	4	>95	$ 6.00[2]
0.25–0.55	Argon	0.51	Glendale Optical Co.	Laser-Gard LGS-A	Goggles	15	45	$50.00[3]
0.60–0.68	HeNe	0.63	Glendale Optical Co.	Laser-Gard LGS-HN	Goggles[6]	5	20	$50.00[3]
0.68–1.30	Ruby	0.69	Fred Reed Optical Co.	BG-18 (Schott filter glass BG-18 in special Glendale frames)[4]	• Plano • Single Correction • Bifocals	5	65	• $77[2] • $86[2] • $105[2]
0.65–0.75	Ruby	—	Glendale Optical Co.	Laser-Gard LGS-R	Goggles	6	20	$50.00[3]
0.90–10.8	Nd : glass HF	1.06 2.7	Fred Reed Optical Co.	KG-3 (Schott filter material KG-3 in special Glendale frames)[4]	• Plano • Single Correction • Bifocals	4.5 3	85 85	• $59[2] • $71[2] • $90[2]
0.75–1.08	Nd	1.06	Glendale Optical Co.	Laser-Gard-LGS-NDGA	Goggles	14	45	$50.00[3]
5.00–10.6	CO_2	10.6	Glendale Optical Co.	Spectacle No. 112A[7]	One piece visitor's spectacles	>10	95	$ 0.50[2]

[1] Table 14-4 adapted from Reference 5.
[2] Prices are July 1, 1975 quotations per unit.
[3] Prices are January 3, 1977 quotations per unit.
[4] Spectacles conform to criteria for industrial mechanical safety at 3 mm thickness and contain pertinent information inscribed on the temples.
[5] Recommendation criteria based on optical density, comfort, visual transmission and cost.
[6] Filters out red light.
[7] Designed as mechanical safety spectacle; has a tendancy to exhibit scratch marks after extended use and should be discarded when visibility is reduced.

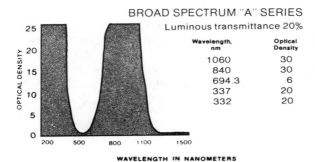

Figure 14-14. Broad spectrum "A" Series. (Courtesy of Glendale Optical Co.)

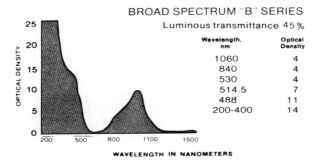

Figure 14-15. Broad spectrum "B" Series. (Courtesy of Glendale Optical Co.)

The Glendale Optical Co. has developed two types of "broad spectrum" safety goggles. See Figures 14-14 and 14-15 for curves showing optical density vs. wavelength. Luminous transmittance values are also included. The unit price as of January 3, 1977 is $90.00 for both the "A" and the "B" Series.

REFERENCES

1. "Laser Protective Eyewear," U.S. Army Environmental Hygiene Agency, Aberdeen Proving Ground, Maryland, 1975.
2. Swope, C. H., "LASERS: Why Eye Protection?," *The Optical Industry and Systems Directory*, 1972–1973.
3. Elder, R. L., "Laser Protective Eyewear," *Applied Optics*, Vol. 13, No. 4, p. 725, April 1974.

4. Industrial Code Rule 50, "Lasers," State of New York, Department of Labor, Board of Standards and Appeals, Albany, N.Y., August 1, 1972.
5. Winburn, D. C., "Selecting Laser-Safety Eyewear," *Electro-Optical Systems Design*, October 1975.
6. "American National Standard for the Safe Use of Lasers, Z 136.1-1976," American National Standards Institute, New York, 1976.

15

Atmospheric Effects

INTRODUCTION

The use of a laser out-of-doors, as for instance on a military-type test range, in a laser communications system or in other applications such as surveying, must sometimes be preceded by calculations to evaluate hazard levels in order to help establish the necessary safety procedures. The calculations should be supplemented by actual measurements. A Class 1 laser would of course present no hazard. Atmospheric attenuation should be considered in the calculations for establishing safe viewing distances for specular or diffuse reflections, and in establishing the attenuation required of safety eyewear.

GENERAL ATMOSPHERIC EFFECTS

Atmospheric attenuation may become a significant factor in determining the radiant exposure or irradiance at ranges greater than a few miles.[1] Atmospheric attenuation is caused by scattering and absorption. Two types of scattering are Mie scattering (by larger aerosol particles) and Rayleigh scattering (by air molecules)[1-3]. Substances that are principal causes of absorption are water vapor, carbon dioxide and ozone.[2] In Rayleigh (or molecular) scattering,

267

the laser light wavelength is much greater than the particle size.[1,3] The scattering is fairly constant at a given wavelength, and is generally the largest factor in the ultraviolet.[1] For the larger aerosols and Mie scattering, the particle size is comparable[3] to or greater[1] than the laser wavelength. Mie scattering is generally the largest factor in the near-infrared and visible spectrum.[1] Gas molecule absorption is frequently a small attenuation factor relative to scattering, and can often be disregarded unless the laser light is in the infrared portion of the spectrum.[1]

Attenuation from scattering is much greater at smaller wavelengths. In much the same way that the red in the sunset still persists after shorter wavelengths from the blue end of the spectrum have faded, the familiar red light from a ruby laser will be scattered significantly less than radiation from the shorter-wavelength portion of the visible spectrum. A clean atmosphere will have relatively less attenuation to the ruby (.6943 micron) and neodymium (1.064 microns) wavelengths. Scattering can attenuate ruby laser radiation by as little as 10% at 10 kilometers and 60% at 100 kilometers. Since the usual value for meteorological visibility is based upon the entire visible spectrum, a quoted visibility value in miles, such as may be obtained from an airport tower, must be used carefully in arriving at the attenuation coefficient for a particular wavelength.[1] Attenuation information for different wavelengths is presented in Figure 15-1.

Atmospheric turbulence causes laser beam scintillation, bringing about localized "hot-spots" within the beam. Scintillation is greatest and causes the largest variations in beam irradiances if air temperature changes relatively quickly with height as compared to a normal atmosphere. Scintillation occurs because of changes in the refractive index of the air along the laser beam path. This results in pointing changes, spreading, refocusing and general power fluctuations over the area of the beam. The effect is strongest in a desert atmosphere with few clouds and weakest on a heavily overcast day. Intrabeam variations can be readily photographed after the beam has traveled a mile or more (the variation of local irradiances within the beam can be a factor of 10 or more). Thus, hot-spots have a probability

Figure 15-1. Approximate variation of attenuation coefficient with wavelength at sea level for various atmospheric conditions. (Note: Neglects absorption by water vapor and carbon dioxide.) (*Courtesy RCA Corp.*)

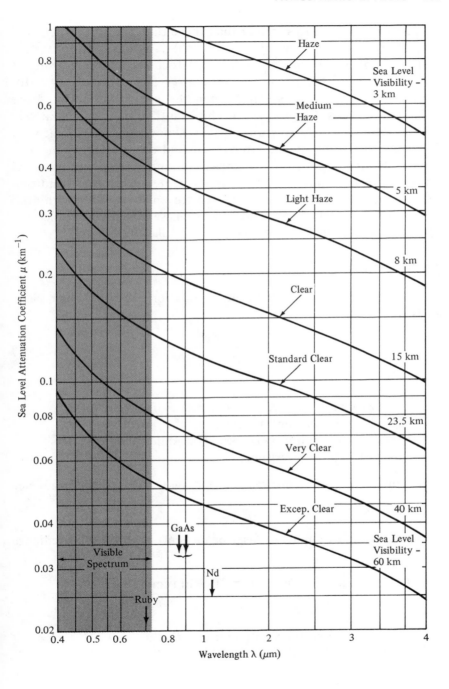

of existing at irradiances greater than those that could be anticipated if the laser beam underwent no atmospheric attenuation.[1] Starting at a range of a few hundred yards, the focusing and defocusing of small portions of the beam can cause a severely irregular beam distribution. Turbulence is greatest when the laser beam path is close to the ground (less than 10 yards). Thus scintillation is comparatively less pronounced for air-to-ground paths. Sliney[4] has offered the view that beam spreading by the atmosphere and other factors can work against the formation of an increased hazard due to hot-spots, as compared to the opinion that hot-spots create a hazardous condition beyond the hazardous range calculated for an atmosphere with little or no turbulence. In any event, a worst-case approach is called for when one is investigating hazards from scintillation effects.

BASIC FORMULA

A principal formula[1,5-7] for hazard evaluation for direct beam viewing is:*

$$I = \frac{1.27 Z e^{-\mu r}}{(a + r\phi)^2} \qquad \text{(Eq. 15-1)}$$

where

I = intensity at range r (H if J · cm^{-2} or E if W · cm^{-2})

r = viewing range in centimeters

Z = level of radiation leaving laser (Q if joules or Φ if watts)

e = base of natural logarithms

ϕ = emergent beam divergence (radians) at $\frac{1}{e}$ × peak-irradiance points

a = diameter of emergent laser beam (cm) at $\frac{1}{e}$ × peak-irradiance points.

μ = atmospheric attenuation coefficient (cm^{-1})

---

*Single-mode laser, Gaussian beam profile; equation accuracy >1% for $\phi < .17$ radian (10°), and >5% for $\phi < .37$ radian (21°). Nonturbulent medium.

If the atmospheric attenuation coefficient is given in km^{-1}, appropriate adjustments must be made to maintain consistent units. Some safety documentation recommends that field evaluations should employ a safety factor of 2 relative to the stated safe levels.[5] Turbulence and scintillation effects should be properly allowed for.

Figure 15-1 gives the atmospheric attenuation coefficient vs. visibility for a number of different wavelengths. Figure 15-2 shows the altitude correction factors used in modifying the atmospheric attenuation coefficient.

Visibility or meteorological range is the horizontal distance at which the typical eye just barely discerns a large black object against

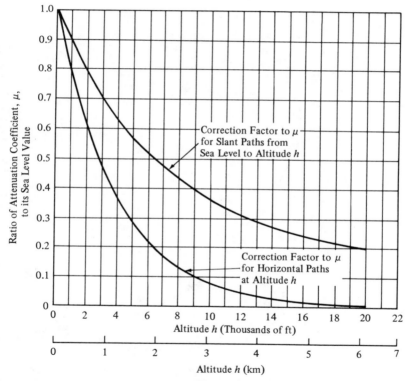

Figure 15-2. Approximate ratio of attenuation coefficient to sea level value for slant paths and horizontal paths. (Note: Neglects absorption by water vapor and carbon dioxide.) (*Courtesy RCA Corp.*)

the horizon sky.[3] To help obtain a measure of visibility more independent of individual eyes, visibility range is defined as the daylight distance at which the contrast is reduced to 2% of the close-up
contrast between the distant black target and the sky.[2] Visibility,
based on the 2% contrast figure, is related to the visible scattering
coefficient, $\mu(\text{km}^{-1})$ as follows:

$$V = 3.9/\mu \qquad \text{(Eq. 15-2)}^{2,3}$$

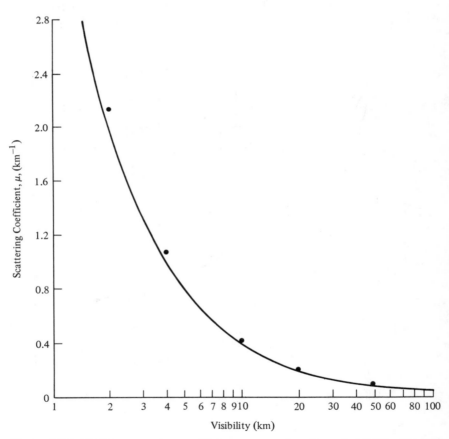

Figure 15-3. Visible scattering coefficient, μ, as a function of visibility. (Notes:
(1) Figure is adapted from Reference 3. (2) Solid points are visual ranges taken
from the international visibility code.)

where

V = visibility (km)

μ = scattering coefficient (km^{-1}) for the visible spectrum

The transmission is related to the scattering coefficient for a given light path by the formula:

$$T = e^{-\mu r}$$ (Eq. 15-3)[2,3]

where

T = transmission over a path of length r (dimensionless)

r = optical path length (km)

μ = scattering coefficient (km^{-1})

Figure 15-3 shows visibility vs. the scattering coefficient.[3] For example, on a day on which the visibility is 10 km, the visible scattering coefficient is approximately .4/km, per the figure. The transmission over a 5-km path is then:

$$T = e^{-.4(5)} = e^{-2.0} = \frac{1}{e^2} = .13, \text{ or } 13\%$$

Figure 15-1 introduces corrections for wavelength and the variation of scattering coefficient with wavelength.

SPECULAR REFLECTIONS

Specular reflection requires a mirrorlike surface. If the reflecting surface is flat, the reflected beam characteristics will be similar to those of the direct beam. The range will be the sum of the distance from the laser source to the reflector plus the distance from the reflector to the eye. If the surface is not flat, for instance a cylindrical or circular object, the reflected power or energy density will be less, and can be calculated if the surface is uniformly curved and the curvature known. Typically, the diameter of the emergent laser beam is many times smaller than the laser beam footprint on the target. Equation 15-1 becomes:

$$I = \frac{1.27 Z e^{-\mu(r + r_1)}}{[(r + r_1)\phi]^2}$$ (Eq. 15-4)

where r_1 = eye distance to target ($a \ll r + r_1$). A worst-case hazard evaluation approach might, for instance, initially assume no atmospheric attenuation.

Reflecting surfaces involved in specular reflections will generally reflect only a fraction of the beam. The magnitude of the reflection is dependent upon the specular reflectivity coefficient and the angle of incidence. For normal (perpendicular) incidence, typical plate glass will reflect only approximately 8% of the incident beam, but at near-grazing incidence, nearly all of the incident radiant energy is reflected.[1]

DIFFUSE REFLECTIONS

Diffuse reflections from a flat surface follow an inverse square law, which once again involves atmospheric attenuation:[5]

$$I = \frac{\rho Z_1}{\pi r_1^2} \quad r_1 \gg D_L \qquad \text{(Eq. 15-5)}$$

where

Z_1 = total energy or power impinging on target ($Z_1 = Ze^{-\mu r}$)
ρ = reflectance of target
r_1 = range from target to point of concern
D_L = diameter of beam at range r

Figure 15-4 shows a representation of some of the symbols used in Eqs. 15-4 and 15-5.

Beyond a few kilometers, atmospheric attenuation becomes an important factor for the target-to-eye distance.

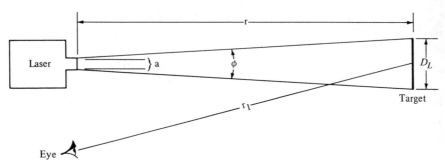

Figure 15-4. Diagram for symbols used, Eqs. 15-4 and 15-5.

If we consider that the laser, the target and the eye are all in the horizontal plane, with the eye along the aim path, then the full effect of the atmosphere can be developed:
Since

$$I = \frac{\rho Z_1}{\pi r_1^2}$$

and with atmospheric attenuation, Z_1 becomes $Z_1 e^{-\mu r_1}$, and $Z_1 = Z e^{-\mu r}$, then

$$I = \frac{\rho Z e^{-\mu r} e^{-\mu r_1}}{\pi r_1^2} = \frac{\rho Z e^{-\mu(r + r_1)}}{\pi r_1^2} \qquad \text{(Eq. 15-5)}$$

Note that $e^{-\mu r} \approx 1$ for small r.

The information in this chapter, presented separately, highlights the role of the atmosphere and its effects on radiant exposure (H) and irradiance (E).

REFERENCES

1. "Control of Hazards to Health from Laser Radiation," U.S. Department of the Army Technical Bulletin TB-MED-279, Washington, D.C., May 1975.
2. *Electro-Optics Handbook*, RCA, Camden, 1966.
3. Wolfe, W. L., *Handbook of Military Infrared Technology*, Office of Naval Research, Department of the Navy, Washington, D.C., 1965.
4. Sliney, D., "The Safety Aspects of Atmospheric Transmission of Lasers," *Annals of the New York Academy of Sciences*, Vol. 267, pp. 366–372, Jan. 30, 1976.
5. NATC Instruction 5100.2, Department of the Navy, Naval Air Test Center, Patuxent River, Maryland, 5 Mar. 1973.
6. "American National Standard for the Safe Use of Lasers, Z 136.1-1976," American National Standards Institute, New York 1976.
7. Sliney, D. H., "Instrumentation and Measurement of Laser Radiation," in "Laser Hazards and Safety in the Military Environment," AGARD Lecture Series No. 79, North Atlantic Treaty Organization, 1975.

BIBLIOGRAPHY

1. Holter, M., et al., *Fundamentals of Infrared Technology*, Macmillan, New York, 1963.
2. Middleton, W. E. K., *Vision Through the Atmosphere*, University of Toronto Press, Toronto, 1958.

APPENDIX *A*

Definitions/Laser Terminology

Absorption—Change of radiant energy to a modified form of energy by interaction with matter; thus a decrease in power of light passing through a substance.

Accessible Emission Level—The magnitude of emission from a laser product of laser or collateral radiation of wavelength and emission duration to which human access is possible as measured pursuant to the applicable sections of the Laser Products Performance Standard.

Accessible Emission Limit—The maximum accessible emission level permitted within a particular laser class as set forth in the Laser Products Performance Standard.

Accommodation—The capability of the eye to modify its power and thus focus for a range of object distances.

ACGIH—American Conference of Government Industrial Hygienists.

Acousto-Optic—Involving action between an acoustic wave and light. Acousto-optic equipment such as modulators and Q-switches is used to control laser light.

Actinic—Of or pertaining to actinism, that is, the intrinsic property in radiation that produces photochemical activity.

Action Spectra Graphs—Graphs that show the comparative spectral ability of actinic radiation to cause reddening of the skin (erythema).

Active Medium—The atomic or molecular substance that can provide gain for laser oscillation. Often called the laser or lasing medium, or active material.

ADA—Ammonium dihydrogen arsenate, $(NH_4)H_2AsO_4$, a crystal used in harmonic generators that has nonlinear optical and electro-optic properties.

ADP—Ammonium dyhydrogen phosphate, $(NH_4)H_2PO_4$, a crystal that has nonlinear optical and electro-optical properties utilized in harmonic generators and modulators; also used in parametric oscillators. The deuterated form $(NH_4)D_2PO_4$ is noted as AD*P.

Advisory Standards—Standards issued by the New York State Department of Labor that are advisory in purpose and intent, and are in no respect intended to be mandatory.

Aerosol—A suspension of minute solid or liquid particles in a gas. The size generally ranges from 100 μm to 0.01 μm, occasionally is less.

Afterimage—An image that remains after a visual stimulus has ceased. Also called "photogene." Can be a symptom of eye injury.

Angstrom (Å)—A unit of length utilized to express length of electromagnetic waves. An angstrom is equal to 10^{-10} meter, 10^{-8} centimeter, 0.1 nanometer or 10^{-4} micrometer (10^{-4} micron).

ANSI—American National Standards Insitute.

Aperture—An opening in the protective housing or other enclosure of a laser product through which laser or collateral radiation is emitted, thereby allowing human access to such radiation.

Aperture Stop—An opening that restricts the size and defines the area across which radiation is measured.

Aspheric—Optical surface that is not a section of a sphere or planar. Used in optical mirrors and lens equipment to avoid spherical aberration.

Atomic Laser—A gas laser in which the medium is of atomic form rather than molecular.

Attenuation—The diminution in the power of an optical beam as it traverses an absorbing and/or scattering substance, as for example the atmosphere.

Audible Emission Indicator—High level tone used to warn personnel of the firing, or imminent firing, of a laser.

Autocollimator—A collimator with a telescope.

Aversion Response—Eye blink; a factor considered important in limiting eye exposure to radiation.

Axial Mode—The mode of frequency of a laser governed primarily by cavity length between the end mirrors, that is, along the cavity length. The mode will have a wavelength which meets the condition that $n\frac{\lambda}{2} = L$, where λ is the laser wavelength, L is the length between the mirrors, and n is some very large integer.

Back Focal Length—The length from a lens's last optical surface to the focal point.

Beam—A group of light rays that can be parallel, or diverge, or converge.

Beam Diameter—The distance between two directly opposite points in a beam at which the power or energy level is a certain fraction, typically $1/e$ or $1/e^2$ of the peak power or energy density of the beam.

Beam Divergence (ϕ)—The growth in beam diameter with range from the exit aperture of the laser. Expressed in millradians at specified points, such as where energy density or power density is $1/e$ or $1/e^2$ times the maximum value.

Beam Expander—A combination of optical parts that will cause the diameter of a laser beam to become larger.

Beam Shutter—Device used to enclose a laser beam without shutting the laser off.

Beam Splitter—An optical part that has reflection properties which can produce two beams from one beam.

Beam Spot Size—Refers to the diameter of the laser beam between the $1/e^2$ power points. Sometimes used loosely to mean the area of the beam cross section between the $1/e^2$ points.

Beam Stop—Utilized to terminate a laser beam at some desired point. Can be an oversized target, densely wooded area, hill, etc. Also called a **backstop**.

Beam Waist—The beam diameter or cross section at the minimum laser beam width. Generally the beam waist is shaped inside of the cavity. Also, if an output divergent beam is later focused by a lens or matched into another cavity, additional beam waists can be shaped.

Bialkali (Na_2KSb)—This substance is utilized as a light-sensitive surface in phototube fabrication.

Birefringent—A crystal having two indices of refraction, often called the ordinary and extraordinary indices. A ray of monochromatic light is divided into two rays with orthogonal polarization by such a crystal. The two rays travel in different directions, except when the input beam travels along a certain direction in the crystal.

Bolometer—A detector measuring infrared light by a temperature-induced resistance variation in a metallic foil exposed to the radiation and heated by it.

Bore Diameter—The inner diameter, such as that of a flash lamp, into which a laser tube or rod is inserted.

Brewster Window—Optical elements that are used to form the ends of a laser cavity, and are used to increase the transmission of one mode of polarization over the other. The windows (transparent glass) are inclined at a specific angle to the laser axis and thus enhance the transmission of a plane polarized light beam. Brewster windows are usually made of fused silica glass.

BRH—Bureau of Radiological Health.

Brightness—The power emitted per unit area per unit solid angle (watts per square centimeter per steradian).

Broadband Spectacles—Laser protective eyewear that gives protection over a broad range of wavelengths.

BSN—Barium sodium niobate ($Ba_2 NaNb_5 O_{15}$), a nonlinear optical material used in the generation of harmonics.

Burst Laser—Term generally used for a pulsed ion laser whose average power and repetition rate can be temporarily increased.

Calorimeter—A detector type that measures the amount of heat energy acquired from a source of electromagnetic radiation, and is used to measure the energy of the source.

Carcinogenic—A cancer-causing agent.

Cardiopulmonary Resuscitation—Methods used to return the heart to a normal beat, after stoppage or irregular beat.

Cataract—Opacity of the lens or cornea of the eye, causing partial or total blindness.

Cavity—The device (resonator) that supplies feedback for laser oscillations. The most common form is comprised of a laser medium between two reflecting surfaces.

CDA—Cesium dihydrogen arsenate ($CsH_2 AsO_4$), a crystal that is used in electro-optic modulators. CD^*A is a different version ($CsD_2 AsO_4$), i.e., containing deuterium rather than hydrogen.

Certificate of Competence—A document issued to the operator of a mobile laser by the Industrial Commissioner in accordance with the provisions of New York State Industrial Code Rule 50.

Certified Mobile Laser Operator—Any individual holding a valid certificate of competence issued by the Industrial Commissioner in accordance with the provisions of New York State Industrial Code Rule 50; the operator is allowed to use a laser outside the laser installation.

Chemical Laser—A laser that obtains population inversion directly by a basic chemical reaction.

Chopping—Modulating or varying the input of a detector at a particular frequency to improve performance.

Choroid—The dark brown vascular coat of the eye between the sclera and the retina.

Ciliary Muscle—The thick part of the vascular coat of the eye that connects the choroid with the iris.

Class I Laser Product—Any laser product that does not permit human access to laser radiation in excess of the accessible emission limits of Class I for any combination of duration and wavelength range.

Class II Laser Product—Any laser product that:

- Permits human access to laser radiation in excess of the accessible emission limits of Class I but not in excess of the accessible emission limits of Class II in the wavelength range greater than 400 nanometers (nm) but less than or equal to 700 nm for emission durations greater than 0.25 second; and,
- Does not permit human access to laser radiation in excess of the accessible emission limits of Class I for any other combination of emission duration and wavelength range.

Class III Laser Product—Any laser product that permits human access to laser radiation in excess of the accessible emission limits of Class I and, if applicable, Class II, but which does not permit human access to laser radiation in excess of the accessible emission limits of Class III for any combination of emission duration and wavelength range. Class III laser products are separately designated as Class IIIa or Class IIIb laser products in labeling, pursuant to applicable sections of the Laser Product Performance Standard.

Class IV Laser Product—Any laser product that permits human access to laser radiation in excess of the accessible emission limits of Class III.

Class V Laser—Extinct classification for an enclosed laser. The American National Standards Institute has eliminated this category in order to make its classification system compatible with that of the Bureau of Radiological Health.

Closed Installation—Any location where lasers are used that will be closed to unprotected personnel during laser operation.

Coherent Light—Radiation in which there is a fixed phase relationship between any two points in the electromagnetic field.

Collateral Radiation—Any electronic product radiation except laser radiation, emitted by a laser product as a result of or necessary for the operation of a laser incorporated into that product. Includes X-radiation.

Collimated Beam—A beam of light where the rays are parallel with very small divergence or convergence.

Collimator—A device for changing a diverging or converging light beam into a collimated or "parallel" beam.

Cone—A cone-shaped detector surface blackened to absorb most of any incoming radiation. The resulting upward change in temperature is measured to determine the radiation level. Also the human eye has a type of photoreceptor in the retina called a cone.

Conjuctiva—The mucous membrane lining the inner eyelid surface and the exposed portion of the eyeball.

Conjunctival Discharge (from eye)—A greater than normal secretion of mucus from the eyeball surface.

Continuous Wave (CW) Laser—A laser whose output lasts for a comparatively long uninterrupted time period, while the laser is actuated (as compared to a pulsed laser). Occasionally meant to mean a laser emitting continuously for a time greater than 0.25 second.

Controlled Area—An area where occupancy and activities are controlled and supervised to give protection from optical radiation hazards.

Cornea—The transparent outside coat of the human eye, which is over the iris. It is a principal refracting part of the eye.

Cryogenics—The scientific and engineering discipline dealing with very low temperatures.

Crystal Laser—A laser whose active medium is an atomic substance in a crystal such as ruby.

Crystalline Lens—The lens of the human eye.

Cycloplegia—Paralysis-induced loss of visual accomodation, due to paralysis of the ciliary muscles of the eye.

Cycloplegic—A substance that brings on paralysis of the eye's ciliary muscle. thereby stopping eye visual accommodation.

Dark Current—A detector's output current, without the presence of incoming radiation, which is due to emission of thermionic electrons. Limits the minimum radiation level that can be detected.

D-CDA—Cesium dihydrogen arsenate (CDA) with dueterium in the place of hydrogen ($CsD_2 AsO_4$). Sometimes abbreviated CD^*A.

Demonstration Laser Product—Any laser product manufactured, designed, intended or promoted for purposes of demonstration, entertainment, advertising display or artistic composition. The term "demonstration laser product" does not apply to laser products that are not manufactured, designed, intended or promoted for such purposes, even though they may be used for those purposes or are intended to demonstrate other applications.

Depigmentation—The removal of the human tissue pigment composed of melanin granules.

Depolarizer—An optical device that causes an unpolarized output, the input beam being polarized.

Dermatology—The medical specialty that deals with skin physiology and pathology.

Designated Individual—An individual selected and directed by a laser safety officer to supervise the operation of a laser.

Detectivity (D*)—A detector sensitivity figure made independent of detector area and bandwidth, in units of centimeter root-hertz per watt. It is the signal-to-noise ratio for the detector with unit incident flux, and changes with incident radiation, temperature of operation and frequency of modulation.

Diffraction Angle—Change in direction of part of a beam, caused by the wave nature of radiation, which occurs as radiation goes past the edge of an opaque obstacle.

Diffraction-Limited—An optical device free of aberrations to a very great extent and limited only by diffraction that unavoidably occurs at the aperture, is called "diffraction-limited." Diffraction-limited optical devices are the "best" that can be achieved. A diffraction-limited spherical lens in conjunction with a circular aperture will focus a uniform intensity monochromatic light beam to a "spot" (Airy disc) of radius $r = (1.22 f \lambda)/D$, where λ equals the wavelength, f equals the focal length of the lens, and D is the lens diameter. A laser beam of Gaussian intensity variation can be focused to a spot radius of $r = (0.64 f \lambda)/D$, where r is the $1/e^2$ power point.

Diffuse Reflection—A reflection of incident beam radiation that reflects energy in many directions from the involved surface or substance.

Diffusion—Variation of beam spatial distribution when the beam is deviated in many directions as it is affected by a surface or medium.

Diode Laser—See "Semiconductor Laser."

DNA—Deoxyribonucleic acid.

Driving Current—For diode lasers, the current required to achieve the lasing operational point; its minimum is the "threshold current" for initiating lasing. The maximum is the amperage that must not be exceeded to prevent diode damage.

Duty-Cycle—Quotient of pulse time duration to repetition period for a repetitively pulsed laser.

Dye Laser—A laser for which the medium is an organic dye, usually in solution, with the solution flowing or contained within a cell. Experimental gas dye and solid lasers have been constructed. Sometimes called organic-dye and tunable-dye lasers.

Edema—Human body tissue swelling caused by abnormal amounts of fluid inside the extracellular spaces.

Electromagnetic Radiation—Energy flow formed by vibrating electric and magnetic fields at right angles and lying transverse to the direction of energy flow. Examples are X-rays, ultraviolet light, visible light, infrared radiation and radio waves, all of which occupy various portions of the electromagnetic spectrum and differ in frequency, wavelength and energy of a quantum.

Electron Beam Excitation—Excitation of a solid state (or a high pressure gas) laser with an electron beam in which the electron energies are higher than 10,000 electron volts.

Electron Beam–Sustained Laser—Molecular gas laser in which the electrical discharge is sustained using a high-energy electron beam. Usually the beam is injected transverse to the laser cavity's optical axis. The electron beam allows laser operation at higher pressures and larger cross section-to-length ratios than those possible with an unsustained discharge. This method is often used in CO_2 lasers with very high continuous wave output power.

Electro-Optic—Term for modulators, Q-switches and other beam control devices that depend on changes in a material's refractive indexes by applying an electrical field. In a Kerr cell, the index variation is proportional to the square of the applied electrical field, with the controlled material generally a liquid. In a pockels cell, the substance is a crystal whose index change varies linearly with the electric field.

Electronic Product—The term is defined in Public Law 90-602 as:
- Any manufactured or assembled product that, when in operation,
  —Contains or acts as part of an electronic circuit and
  —Emits (or in the absence of effective shielding or other controls would emit) electronic product radiation;

OR

- Any manufactured or assembled article that is intended for use as a component, part or accessory of a product described above and which when in operation emits (or in the absence of effective shielding or other controls would emit) such radiation.

Electronic Product Radiation—The term is defined in Public Law 90-602 as:
- Any ionizing or nonionizing electromagnetic or particulate radiation;

OR

- Any sonic, infrasonic or ultrasonic wave which is emitted from an electronic product as the result of the operation of an electronic circuit in such product.

Emergent Beam Diameter—Diameter of the laser beam where it exits from the output aperture of the laser system.

Emission Duration—The temporal duration of a pulse, a series of pulses or continuous operation, expressed in seconds, during which human access to laser or collateral radiation could be permitted as a result of operation, maintenance or service of a laser product.

Enabling Legislation—Authorization by a state to its respective state health department or equivalent agencies to adopt some form of laser safety standards.

Enclosed Laser System—Any laser or laser system located within an enclosure that does not permit emission of hazardous optical radiation from the enclosure.

Energy (Q)—The capacity for doing work. Energy content is commonly used to characterize the output from pulsed lasers, and is generally measured in joules (J).

Energy Density—The energy per unit area expressed in units of joules per square centimeter (J/cm^2). Same as "Radiant Exposure."

Epithelium (of the Cornea)—The layer of cells forming the surface or skin of the cornea.

Erythema—A redness of the skin tissue due to capillary congestion. A skin burn effect caused by overexposure to a laser beam.

Etalon—An optical element which can be placed in a laser's cavity to restrict the output bandwidth. Etalons are utilized widely in dye lasers when the intended use calls for a narrow bandwidth output beam.

Exposure—The product of irradiance and the time it lasts.

Extended Source—A source of radiation that can be resolved by the eye to form a geometrical image, as compared to a point source, which the eye cannot resolve into a geometrical image.

Fail Safe—A term denoting a device or system that can compensate automatically for a failure.

Far-Field—That vicinity which is beyond a diffracting aperture by a length of about 20 to 100 times the quotient $\dfrac{\text{aperture area}}{\text{wavelength}}$. For a laser with an output aperture area of 10^{-2} cm^2 and with a wavelength of 5,000 Å (5×10^{-5} cm), the far-field can be considered to be beyond $100 \times 10^{-2}/5 \times 10^{-5}$ or 2×10^4 cm (200 meters). When a laser beam is tested in the far-field, all test results acquired using Fraunhofer diffraction principles are valid.

Faraday Effect—Rotation of the polarization plane of a light beam that occurs when the beam passes through a homogeneous substance in the direction of an applied magnetic field. The extent of the rotation varies with the material, distance traveled and the strength of the magnetic field.

Faraday Rotator—An optical device for rotating the polarization plane of a laser beam using the Faraday effect.

Filter Damage Threshold—The radiation energy or power level from a laser that can cause protective eyewear filter material to be damaged (bleach, bubble, melt or shatter).

Focal Length—The length between the secondary nodal point of a lens and the primary focal point. For a thin lens, the focal length is the length between the lens and the focal point.

Focal Point—The focus point toward which light waves converge or from which they diverge or appear to diverge.

Fovea Centralis—A small depressed spot in the macula lutea of the retina, which is the area of clearest vision.

Frequency Multiplier—A nonlinear element that generates a multiple of the input frequency (for instance, a frequency doubler). Sometimes called a **harmonic generator.**

Fresnel Lens—A lens with the face broken up into many concentric annular rings that combine to provide the same focusing power as a simple lens. Fresnel lenses have the advantage of being thinner than simple lenses, and are often used for short focal length purposes to prevent excessive thickness.

Gas Laser—A laser for which the medium is a gas, of either atomic or molecular form. This laser type is subdivided by medium into atomic (such as helium-neon), molecular (such as carbon dioxide) and ionic (that is, argon, krypton, xenon and other types such as the metal-vapor helium-cadmium laser). Frequently, "ion" is taken to mean argon or krypton.

Gaussian Distribution—A frequency distribution curve for a population or collection of variable data, frequently shown as a bell-shaped curve symmetrical about the mean of the data. Frequently called **normal distribution.**

Glass Laser—A solid state laser in which the active medium consists of a rare earth atomic species contained in a glass rod.

Golay Cell—A broadband sensor of radiation energy, which functions by measuring the rise of pressure of a gas held in an illuminated, transparent cell. The rise in pressure varies with the temperature change caused by the incident radiation.

Harmonic Generator—A nonlinear element that produces a harmonic of the frequency of the original input laser beam. Frequently called a **frequency multiplier.**

Hertz (Hz)—The frequency of a periodic oscillation, given in cycles per second.

HEW—Department of Health, Education and Welfare.

Hot Spots—Localized higher-intensity parts of a laser beam, frequently caused by atmospheric turbulence effects.

Human Access—Access at a particular point to laser or collateral radiation by any part of the human body, by a straight line having an unobstructed length of 100 centimeters, or by any line having an unobstructed length of 10 centimeters, when laser or collateral radiation is incident at that point.

Infrared Radiation—Electromagnetic radiation the wavelengths for which are within the spectrum range of 0.7 μm to 1 mm. This portion of the spectrum is often separated into three bands by wavelength: IR-A (0.7–0.78 μm to 1.4 μm), or near-infrared; IR-B (1.4 μm to 3 μm); and IR-C (3 μm to 1 mm), or far-infrared.

Injection Laser—Refer to "Semiconductor Laser."

Inorganic Liquid Laser—A laser for which the active medium substance is an inorganic liquid.

Integrated Radiance—The radiant energy per unit area of a radiating surface per unit solid angle of emission, stated in joules per square centimeter per steradian ($J \cdot cm^{-2} \cdot sr^{-1}$).

Intrabeam Viewing—Generally speaking, viewing the laser source from within the beam. The beam may either be direct or specularly reflected.

Ion Laser—A laser for which the active medium is an ionized gas, such as argon or krypton.

Ionizing Radiation—Radiation that can produce ionization, such as sufficiently energetic charged particles such as alpha and beta rays, and wave-type radiation such as X-rays.

Iris—The circular colored membrane that encircles the pupil and is positioned behind the cornea of the eye.

Irradiance (E)—The radiant power that falls on a surface divided by the area irradiated, expressed in watts per square centimeter ($W \cdot cm^{-2}$).

Joule (J)—A unit of energy; one (1) joule = 1 watt for one second (watt·second) or 10^7 ergs or 0.239 calorie.

Joule · cm^{-2} ($J \circ cm^{-2}$)—A unit of radiant exposure utilized when measuring the amount of energy per unit area of surface or per unit area of a laser beam.

k_1 and k_2—Wavelength and sampling interval–dependent correction factors for Class I, II, III and IV laser Accessible Emission Limit tables, per BRH Laser Products Performance Standards.

KB5—Potassium pentaborate ($KB_5O_8 \cdot 4H_2O$), a nonlinear crystal used for phase matched harmonic generation at short wavelengths. Also called **KPB**.

KDP—Potassium dihydrogen phosphate (KH_2PO_4), a crystal type substance with nonlinear optical properties that can be used in electro-optic modulators. The deuterium-containing form of the crystal (KD_2PO_4) is referred to as D-KDP or KD*P.

Kerr Cell—A beam modulator that uses the Kerr effect. The extent of the modulation varies with the square of the applied electric field. See "Electro-Optic."

Key Actuated Master Control—Safety feature frequently designed in with Class III and Class IV lasers, which prohibits firing the laser until the key is used.

Krypton Tube—A gas-filled cold cathode switching tube that passes high peak currents for brief intervals while operating in an arc discharge manner.

Lambertian Surface—A perfectly diffusing surface, whose emitted or reflected radiance (brightness) is not dependent on the viewing angle.

Laser—A word for light amplification by stimulated emission of radiation. Lasers generate or amplify electromagnetic oscillations at wavelengths from the far infrared (submillimeter) to the ultraviolet. The laser oscillator needs two basic elements: an amplifying medium and a regeneration or feedback mechanism (resonant cavity). The amplifying medium can be a variety of substances, such as a gas, semiconductor, dye solution, etc. Feedback is generally created by two mirrors. The distinctive properties of the resulting electromagnetic oscillations include monochromaticity, extremely high intensity, very small bandwidth, very tight beam divergence and phase coherence.

Laser Amplifier—A device that increases the energy of a pulse produced by a laser oscillator. The active medium is pumped in the same manner as in a basic laser oscillator; however, the amplifier does not contain the resonant cavity needed to originally generate pulses.

Laser Beam Width ($1/e^2$ Power-Points)—The power in a laser beam, instead of going to zero at a precise radius from the center of the beam, fades away slowly, going to zero at infinity. Because of this, the beam width is frequently defined as the diameter in between the points at which the beam intensity is $1/e^2$ of its value at beam center. The $1/e$ point, as well as the half power point, is occasionally used for defining the beam diameter.

Laser Cavity (Optical or Resonant Cavity)—The enclosed volume (resonator) that enables feedback for laser oscillation and light generation. Frequently, the configuration is formed of two reflecting mirrors, separated by the cavity length, L. The lasing medium is positioned between the end mirrors.

Laser Classification—A ranking of comparative beam hazard, which is of great importance in setting the necessary control measures for reducing the radiation hazard. Four classifications exist (Class I, II, III and IV).

Laser Controlled Area—Any area in which there are one or more lasers, and in which the personnel activity and laser utilized are under control and supervision; also, entry to the area may be controlled.

Laser Energy Source—Any piece of apparatus or equipment meant for use with a laser to provide energy for the functioning of the laser. Usual energy sources such as electrical supply mains or batteries are not to be considered as laser energy sources, per the BRH laser Products Performance Standard.

Laser Footprint—The laser beam pattern or distribution as seen and/or recorded by video camera and recorder or other techniques (e.g., photographic).

Laser Head—The enclosure for the active medium, cavity and other parts of a laser except for the power supply.

Laser Installation—Per New York State Industrial Code Rule 50, any location where for a period of more than 30 days one or more lasers are used or operated. The confines of a laser installation shall be designated by the owner of such installation. An entire building or other structure, a part thereof or a plant may be designated as a laser installation. A construction site shall not be considered a laser installation.

Laser Medium—See "Active Medium."

Laser Product—Any manufactured product or assemblage of components that constitutes, incorporates or is intended to incorporate a laser or laser system. A laser or laser system that is intended for use as a component of an electronic product shall itself be considered a laser product.

Laser Pumping—The activation of a laser medium by absorption of energy. Laser mediums can be "pumped" in many different ways—with flash lamps, electron beams, etc. Electron or molecular transitions take place to higher energy levels.

Laser Radiation—All electromagnetic radiation emitted by a laser product within the spectral range of approximately 250 nm to 13,000 nm, which is produced as a result of controlled stimulated emission, or which is detectable with radiation so produced within the appropriate aperture stop specified in the Laser Products Performance Standards.

Laser Safety Officer—An individual designated at a particular laser installation or for a particular mobile laser, who is qualified by training and experience in the occupational and public health aspects of lasers to evaluate the radiation hazards of such laser installation or mobile laser, and who is qualified to establish and administer a laser radiation protection program for such laser installation or mobile laser, and has definite authority for supervision of the control of laser hazards.

Laser Safety Program—The steps taken to organize and develop fully a program for controlling laser hazards.

Light—Visible radiation (400 nm to 700 nm).

Limiting Angular Subtense (α_{min})—The visual angle that separates intrabeam viewing from extended source viewing.

Limiting Aperture—The maximum circular opening through which radiance and radiant exposure measurements can be averaged.

Linewidth—The frequency range over which the greater part of the energy of the laser is distributed. Can also be referred to as **bandwidth**. In dye lasers, frequently called **spectral bandwidth.**

Liquid Laser—A laser that has an active element of either an organic dye or inorganic liquid.

Lossy Medium—A medium that attenuates (absorbs or scatters) radiation which passes through it.

Luminous Transmittance—Term by which an individual's ability to see through an optical filter (such as a protective lens) can be expressed; frequently given as a percent.

Maintenance—Performance of those adjustments or procedures specified in user information provided by the manufacturer with the laser product, which are to be performed by the user for the purpose of assuring the intended performance of the product. It does not include "operation" or "service" as defined in the Laser Products Performance Standards (and in this Appendix).

Maser—A device for radio microwave amplification by the stimulated emission of radiation. In contrast to a laser which emits light, a maser emits microwave radiation.

Maser (Optical)—Another, older and now seldom used name for a laser, with the term microwave (of optical frequency) replacing light in laser.

Maximum Output—The maximum radiant power and, where applicable, the maximum radiant energy per pulse of accessible laser radiation emitted by a laser product during operation, as determined pursuant to the applicable section(s) of the Laser Products Performance Standards.

Maximum Permissible Energy Density for Corneal and Skin Exposure—The energy density of laser radiation from a pulsed laser that, in accordance with present medical knowledge, is not expected to cause detectable corneal or other bodily injury to an individual at any time during his lifetime.

Maximum Permissible Power Density for Corneal and Skin Exposure—The power density of laser radiation from a CW or pulsed laser that, in accordance with present medical knowledge, is not expected to cause detectable corneal or other bodily injury to an individual at any time during his lifetime.

Medical Laser Product—Any laser product manufactured, designed, intended or promoted for purposes of *in vivo* diagnostic, surgical or therapeutic laser irradiation of any part of the human body.

Melanin—A dark pigment that is in the skin, retina and hair.

Metal Vapor Laser—A laser that has as the active element a metal vapor, usually in combination with another gas, such as helium-selenium or helium-cadmium.

Meter (m)—A standard unit of length in the international metric system of units. The meter is subdivided into the following units:

Centimeter	$cm = 10^{-2} m$
Millimeter	$mm = 10^{-3} m$
Micrometer	$\mu m = 10^{-6} m$
Nanometer	$nm = 10^{-9} m$

Micrometer (μm)—A length equal to 10^{-6} meter or 1,000 nanometers, often termed **micron**.

Mixed Gas Laser—As sometimes used, refers to an ion laser with a combination of argon and krypton.

Mobile Laser—A laser that is used or operated outside a laser installation.

Mode-locking—Internal modulation of the laser at a frequency that equals the natural distance between the modes of the laser resonator, which results in a burst of short pulses at the modulation frequency with pulsewidth approximately equal to the reciprocal of the laser's natural line width.

Mode-Locked Laser—A laser for which the phase or amplitude of the laser's output is locked so that the output leaves the cavity in a controlled manner. The output consists of a burst of very short pulses. By way of contrast, the unlocked, free-running modes in a laser cavity cause low-frequency "noise" output because of random phasing and time-varying amplitudes in the cavity. Sometimes called **phase-locked**.

Molecular Laser—A laser that has as the active element a molecular gas (carbon dioxide or water vapor are examples).

Monochromaticity—The condition of containing or generating light of only one wavelength, as for instance by a laser.

Monochromator—A device that selects light radiation in a narrow band of wavelengths from a range of wavelengths present in a beam.

MPE—Maximum Permissible Exposure.

Multialkali ([Cs]Na$_2$KSb)—A substance that is utilized as a light sensitive surface in phototubes.

Multiline—Used in describing ion lasers, denotes simultaneous emission at greater than one wavelength. In molecular lasers such as CO_2, multiline denotes simultaneous emission from greater than one rotational-vibrational change or transition of the CO_2 molecule.

Multimode—Output or emission at several frequencies simultaneously, usually closely spaced, where each frequency represents a different mode of laser oscillation in the cavity.

Mydriatic—A drug that produces dilation or spreading of the pupils.

Nanometer (nm)—A unit of length equal to 10^{-9} meter, 10^{-7} centimeter, 10^{-3} micron or 10 Angstroms. The nanometer is beginning to replace the Angstrom as the principal unit of wavelength measurement of electromagnetic wavelengths.

Nd-Glass—Neodymium doped glass, which is utilized in some solid state lasers. The neodymium atoms form the active medium.

Nd-YAG—Neodymium doped yttrium aluminum garnet, which is a crystal type utilized in a common type of solid state laser. The neodymium atoms form the active medium.

Near-Field—That vicinity which is between the diffraction aperture and a length given by the quotient of $\dfrac{\text{aperture area}}{\text{wavelength}}$. Therefore, for a laser which has an output aperture area of 10^{-2} cm^2 and a wavelength of 5,000 Å (5×10^{-5} cm), the near-field would be the distance within 2 meters of the laser. The distance between 2 meters and, say, 40 meters is a "gray" area. While it is usually considered to be in the near-field, in order to be conservative in some instances, where accuracy is not important, it can be considered the far-field. When a beam is measured and analyzed in the near-field, Fresnel diffraction principles need to be applied to the beam parameters.

Noise Equivalent Power (NEP)—Denotes the minimum power which a detector can reliably record, usually given for a particular wavelength, frequency and noise bandwidth. Usually expressed in watts. NEP is rigorously expressed as the ratio of the root mean square noise signal (volts or amperes) to the detector's responsivity or sensitivity (in volts or amperes per watt).

Nonlinear Effects—Variations in a medium generating electromagnetic waves which vary as the second, third or higher powers of the external electric field. These nonlinear optical effects include harmonic generation.

Ocular—Relating to the eye.

Ocular Fundus—The inner base surface of the eye, most removed from the opening of the eye, the part opposite the pupil.

Open Installation—Any location where lasers are used that will be open to operating personnel during laser operation and may or may not specifically restrict casual entry.

Operation—The performance of the laser product over the full range of its functions. It does not include "maintenance" or "service" as defined in the Laser Products Performance Standards (and in this Appendix).

Ophthalmology—The medical specialty that includes eye anatomy, functions, pathology and treatment.

Ophthalmoscope—A device basically made of a mirror that has a central opening through which the eye is examined.

Optic Nerve—The sensory nerve of each eye that connects the retina to the brain.

Optical Cavity—See "Cavity."

Optical Density (O.D.)—A number that equals the common logarithm (base 10) of the attenuation afforded by a filter or other attenuating medium:

$$\text{Optical density: O.D.} = \log_{10}[\Phi_0/\Phi_t]$$

where Φ_0 is the input power and Φ_t is the remaining power at a specific wavelength λ. Also, the common logarithm of the inverse of the transmittance:

$$\text{O.D.} = -\log_{10}\tau_\lambda \quad (\tau \text{ is transmittance})$$

Optically Pumped Laser—A laser that has an active medium which is excited (pumped) with another light source to cause population inversion. For solid state lasers and for some dye lasers, this pump source generally is an incoherent type (i.e., flash lamp or arc lamp). For some gas and other dye lasers, coherent laser sources frequently provide the optical pumping.

Organic Dye Laser—See "Dye Laser."

Oscillator-Amplifier—A device used for producing very large laser powers, frequently a gigawatt or more. The laser oscillator output is amplified by one or more follow-on external laser amplifiers.

OSHA—Occupational Safety and Health Administration.

Output Power and Output Energy—Output power is used frequently to rate CW lasers where the energy delivered per unit time (watts) stays relatively constant. Pulsed lasers are often categorized by energy output per pulse.

P_{exempt}—The output power (Q_{exempt}, the output energy per pulse) for a laser of a level where no relevant Maximum Permissible Exposure for the eye can be exceeded, whether or not optical instruments are used.

Parametric Oscillator—A nonlinear device, frequently a crystal, that provides tunable laser oscillations equal to the sum or difference frequency of mixed electromagnetic waves. Sometimes referred to as a tunable parametric oscillator or as an optical parametric oscillator; loosely used to denote the complete instrument consisting of the laser and the tuning crystal.

Pathology—The study of disease, its nature, causes, processes, development and consequences. Also denotes the anatomic or functional symptoms of disease.

Photometer—An instrument or device used to measure or evaluate light intensity. Rigorously speaking, a photometer measures light intensity in photometric units (lumens), which correspond to the intensity of light that is perceived by the human eye. However, the word has been used for systems that measured intensity in watts at frequencies outside the visible region.

Photon—The quantum of electromagnetic energy, equal to the product of Planck's constant and the frequency of the radiation.

Photon Drag Detector—A type of detector that, as infrared radiation passes through a special bar of doped germanium and reacts with its free carriers, produces a voltage drop across the crystal which varies with the power of the input beam. The fast interaction allows detection of infrared radiation using a room-temperature device.

Photophobia—Aversion to light, frequently caused by physical discomfort resulting from exposure to strong light.

Photosensitizers—Materials that are used to increase sensitivity of substances to irradiation by electromagnetic radiation.

Photovoltaic—A device capable of generating a voltage when incident radiant energy, especially light, falls on it.

Pigment Epithelium (of the Retina)—The layer or lining of cells that have as a component brown or black pigment granules adjacent to and behind the rods and cone receptor cells.

PIN—A type of semiconductor detector that contains an intrinsic semiconductor region in between the p- and n- regions.

Pockels Cell—A modulator that utilizes the Pockels effect, in which the modulation varies linearly with the applied electric field; also used as Q-switch.

Photoconductivity—The term used to describe the change in electrical conductivity that light causes in some substances.

Photoemission—The emission of photoelectrons, particularly from metallic surfaces.

Point Source—A source of radiation whose dimensions are quite small compared with the viewing distance, or at least minute enough in comparison with the distance between source and detector for them to be neglected in calculations.

Population Inversion—A situation created in laser mediums by pumping where a higher energy state has more excited atoms compared to at least one lower energy state. Normally, higher energy states are less populated than the accompanying lower energy states.

Power (Φ)—The time rate at which energy is given off, received, or used in some fashion; expressed in watts or in joules per second.

Power, Average—With a pulsed laser, the pulse energy (joules) times the frequency of the pulse (Hertz), expressed in watts; as compared to the average power of a pulse.

Power Density—A phrase used to denote the power per unit area ($W \cdot cm^{-2}$, for example) contained in a laser beam, or falling on a given target area. It is the same as the radiometric term **irradiance.**

PRF—Pulse repetition frequency. The frequency of laser pulses, in pulses per second.

Protective Housing—Those portions of a laser product that are designed to prevent human access to laser or collateral radiation in excess of the prescribed accessible emission limits under conditions specified in the applicable sections of the Laser Products Performance Standards.

Proustite ($Ag_3 AsS_3$)—A nonlinear crystal utilized for harmonic generation that also manifests the electro-optic effect (see "Electro-Optic").

Pulse Duration—The time delta measured between the half-peak-power points at the front and trailing edges of a pulse (pulsewidth).

Pulse Selector—A device that picks one optical pulse from a burst of mode-locked pulses. Similar devices are available for choosing an electrical pulse from the output of an electronic pulse generator.

Pulsed Laser—A laser that generates energy in pulses rather than continuously. For some purposes, where the pulse duration is ≤ 0.25 second.

Pump—The laser energy source (such as a flash lamp, electron beam or current supply) necessary for amplification in the active medium of a laser.

Pupil—The variable opening in the iris through which light passes toward the interior regions of the eye.

Pyroelectric—A crystal that manifests electrical effects when its temperature is changed. Generally considered a current source that has an output varying with the rate of change of temperature. Used to detect and measure infrared radiation.

Pyrometer—A device or instrument utilized in the measurement of high temperatures, sometimes remotely.

Q-switch—A device that exhibits a shutter type effect which prevents laser emission until it is opened. "Q" denotes the quality factor of the laser's resonant cavity. "Active" Q-switching can be accomplished with a Kerr or pockels cell, a rotating mirror, rotating prism, or acousto-optic devices. "Passive" Q-switching is accomplished using a saturable absorber substance such as a gas or a dye. In a pulsed laser, a Q-switch greatly increases pulse power by lessening pulse duration while maintaining the energy constant.

Radian—A unit of angular measure equal to the angle subtended at the center of a circle by a chord whose length is equal to the radius of the circle. One (1) radiant ≈ 57.3 degrees; 2π radians = 360 degrees.

Radiance (L)—The radiant power per unit area of a radiating surface per unit solid angle of emission, expressed in watts per square centimeter per steradian ($W \cdot cm^{-2} \cdot sr^{-1}$).

Radiant Energy (Q)—The energy given off, received or used in some manner, in the form of radiation, expressed in joules (J).

Radiant Exposure (H)—The radiant energy irradiating a portion of a surface divided by the area of that portion, in joules per square centimerer ($J \cdot cm^{-2}$).

Radiant Flux—See "Radiant Power."

Radiant Power (Φ)—Power given off, received or used in some fashion, in the form of radiation; the time rate of transfer of radiant energy. Expressed in watts (W). Also called radiant flux.

Radiant Intensity (I)—Radiant power of a source in a particular direction; ratio of the radiant flux leaving the source, transmittted in a portion of the solid angle containing the given direction, and the portion of the solid angle. Units: watts per steradian (W · sr^{-1}).

Radiometer—A device or instrument utilized to measure radiation in watts. Radiometer measurements can be made at any wavelength. The useful spectral range of a particular device may be confined to a narrow range.

Radiometric Units—A set of units used in the measurement of the intensity of electromagnetic radiation. One basic unit is the watt.

Rayleigh Scattering—Scattering of radiation as it passes through a medium containing particles; in Rayleigh scattering, the particle sizes are small relative to the wavelength of the radiation.

RDA—Rubidium dihydrogen arsenate (RbH_2AsO_4), a type of crystal substance with nonlinear optical properties used in harmonic generation.

Receiver—A unit that is comprised of a detector and signal handling electronics which changes optical input into electronic output. Frequently used in communications.

Reflectance, Reflectivity (ρ)—The ratio of total reflected radiant power or flux to total incident power or flux.

Reflection—The casting back, turning back or deviation of radiation following its impacting a surface.

Refraction—The deflection or bending of a wave, as of light or sound, at the boundary between two mediums that have different refractive indices, or as the waves pass through a substance of nonuniform density.

Remote Control Connector—An electrical connector that permits the connection of external controls placed apart from other components of the laser product to prevent human access to all laser and collateral radiation in excess of the limits specified in the applicable sections of the Laser Products Performance Standards.

Removal Laser System—Any laser system that is incorporated into a laser product and which is capable, without modification, of producing laser radiation when removed from such laser product, shall itself be considered a laser product, and therefore subject to the provisions of the Laser Products Performance Standards.

Repetitively Pulsed, Q-switched Lasers—A term that denotes lasers whose energy is generated in pulses under control of a Q-switch.

Research and Development—Defined in New York State Industrial Code Rule 50 as theoretical analysis, exploration or experimentation. Also, the extension of investigative findings and theories of a scientific or technical nature into practical application for experimental, demonstrative and specialized purposes including the experimental or limited production and testing of models, devices, equipment, materials and processes involving the use of lasers.

Resonant Cavity—See "Cavity."

Retina—The sensory membrane that receives an image transmitted and shaped by the cornea and lens of the human eye, and which lines much of the inside portion of the eye.

Retroreflector—A prism or mirror or device that reflects incident rays in such a manner that the outgoing beam is along the same line as the input beam but travels in the opposite direction.

RNA—Ribonucleic acid.

Rods—A rodlike type cell in the retina that is sensitive to dim light.

S1 to S24—Standard designations for light-sensitive surfaces in phototubes.

Safety Interlock—A device associated with the protective housing of a laser product to prevent human access to excessive radiation in accordance with the applicable sections of the Laser Products Performance Standards.

Sampling Interval—The time interval during which the level of accessible laser or collateral radiation is sampled by a measurement process. The magnitude of the sampling interval in units of seconds is represented by the symbol t.

Scanned Laser Radiation—Laser beam emissions having a time-varying path, source or pattern of transmission with respect to a fixed frame of reference.

Scintillation—A term utilized to describe the quick variations in irradiance levels, caused by atmospheric turbulence, in a cross section of a laser beam.

Sclera—A tissue that is the tough, white, fibrous outside envelope which covers most of the eyeball except the cornea.

Semiconductor Laser—A laser formed from an active material that is a semiconductor, which can be a diode or homogeneous. Commercial types are usually diodes in which lasing takes place at the junction of n- and p-type semiconductors, typically gallium aluminum arsenide or gallium arsenide. Homogeneous types are composed of undoped semiconductor material and are energized (pumped) by an electron beam.

Sensitivity—The smallest input that can be measured by a detector type device, or the ratio of the device's output signal to its input signal.

Service—The performance of those procedures or adjustments described in the manufacturer's service instructions that may affect any aspect of the product's performance. It does not include "maintenance" or "operation" as defined in the Laser Products Performance Standards (and in this Appendix).

Shield—A specially built enclosure for a laser, a laser beam and/or a target to prevent hazardous emissions. Laser shielding materials are frequently made of rigid plastics, metals and other materials such as opaque ceramics and light-impervious cloth. The term is used to denote a radiation shield or an explosion shield, or else a combination radiation and explosion shield.

Signal Averager—An electronic device that outputs the mean or average of a number of pulses from a pulse source.

Single-Mode Operation—Laser output at a single TEM mode. If the single mode is TEM_{00}, the laser provides a single spot of beamed light that spreads only a minimum amount during transmission and is capable of being focused to the smallest spot size.

"Smart" Bomb—A laser-guided bomb as compared to a conventional ballistic falling bomb.

Soleil-Babinet Compensator—A type of retardation plate that can be varied continuously to produce a desired degree of retardation.

Solid Angle (ω)—The ratio of a particular area on the surface of a sphere to the square of the sphere radius. Expressed in steradians (sr).

Solid State Laser—A laser in which the active medium is an atomic substance in a glass or crystal. The atomic substance can be deliberately added to the crystal or glass (for instance, neodymium is added to glass), or may be intrinsic, such as the familiar case of chromium in ruby. This term is generally not applied to semiconductor lasers.

Source—This term can mean either a laser or a laser-illuminated reflecting surface.

Spatial Filtering—An optical technique that involves filtering diffraction patterns and thus re-forms a desired image. The device used is termed a spatial filter since it is indeed a space filter, being a mask with a specific pattern of holes or grooves cut out. By positioning a spatial filter at a particular position in an optical device, the blurring effect of dust particles on a certain image (noise) can be efficiently filtered out.

Spectral Range—Term used with different meanings for detectors and dye lasers. For a detector, it denotes wavelengths of radiation to which the detector responds. When used, for dye lasers, it denotes the wavelength range through which the output can be tuned.

Specular Reflection—A mirrorlike reflection.

Steradian (sr)—The unit of measure for a solid angle. There are 4π steradians in a sphere.

Stimulated Emission—Emission (radiation) of electromagnetic energy during a transition from a higher energy state to a lower energy state in an activated laser medium. The emission is made possible by the presence of a radiation field (stimulating radiation). The stimulated emission is practically the same in every way (frequency, wavelength, momentum, phase, polarization) compared to the stimulating radiation.

Streak Camera—A very high-speed camera device in which an image of a line in an occurrence is moved across a photosensitive plate at a constant rate, thus providing a record of light intensity as a time variable.

Superradiant—Coherent optical amplification of spontaneous emission that takes place without relaxation processes. It is generally used to denote a laser whose gain is high enough to allow amplification without mirrors. Two examples are nitrogen and molecular hydrogen. The inherent beam quality of a superradiant laser is generally poor compared to that of a laser with a complete optical cavity. "Superfluorescent" has been suggested as a more rigorous term for this type of laser.

Survey—An evaluation of the laser radiation hazards incident to the production, use, disposal, servicing or presence of any laser including, if appropriate, a physical evaluation of the laser installation or mobile laser and the measurement of laser output and reflected laser radiation.

Surveying, Leveling or Alignment Laser Product—A laser product manufactured, designed, intended or promoted for one or more of the following uses:

- Determining and delineating the form, extent or position of a point, body or area by taking angular measurement.
- Positioning or adjusting parts in proper relation to one another.
- Defining a plane, level, elevation or straight line.

TEA Laser—An acronym for a transversely excited, atmospheric pressure laser. A TEA laser is a gas type laser for which energizing or excitation of the active medium is provided transverse to the flow path of the medium. Because of the shorter breakdown length, the TEA laser functions in a gas pressure range above that for longitudinally excited gas lasers (but not necessarily atmospheric). TEA lasers provide potentially higher power output per unit volume because of the higher concentration of lasing molecules per unit volume.

TEM$_{00}$—The radiation from a laser operating in the fundamental transverse mode. The energy possesses a Gaussian (bell-shaped) distribution. All the emitted energy is in one spot, with no side lobes.

TGS—Triglycine sulfate, a nonlinear material.

Thermal Imaging Plate—A plate that has a surface sensitive to thermal effects, which provides a visual display of infrared laser beams.

Thermocouple—A thermoelectric sensor utilized to measure temperature accurately, particularly one made of two dissimilar metals joined so that a voltage difference is produced between the points of contact and indicates temperature.

Thermoelectric—Term for electrical phenomena that occur in conjunction with a flow of heat.

Thyratron—A gas-filled hot-cathode tube utilized to switch high currents.

Time Delay Generator—A device that provides an electronic pulse at a preset interval subsequent to an initial pulse. Sometimes called a **digital delay generator**.

Transmission—Passage of electromagnetic energy through a medium.

Transmittance or Transmissivity (τ)--The ratio of total transmitted radiant power to total incident radiant power.

Transverse Mode—Refers to the change in intensity of laser light in the cross section of the wavefront of a laser beam. Many different transverse modes are possible (TEM_{00}, TEM_{01}, TEM_{10}, TEM_{11}, etc.), but a laser can operate preferentially in one or more of the permitted transverse modes. A laser can be forced to operate in a given mode (TEM_{00}, for example) by varying the energy loss/gain of the cavity to favor operation in the desired mode.

Tunable Laser—A laser that allows continuous variation of laser emission across a broad wavelength region. A single laser system can be "tuned" to radiate laser light over a continuous spectrum of wavelengths or frequencies.

Ultraviolet Radiation—Electromagnetic radiation that has wavelengths from soft X-rays down to visible, violet light. This spectral region is frequently categorized into three separate bands by wavelength: UV-A (315–400 nm), UV-B (280–315 nm) and UV-C (200–280 nm). Ultraviolet wavelength is shorter than wavelength for visible radiation.

Ultraviolet (UV) Preionized Laser—A pulsed gas laser in which "seed" electrons produced by a preliminary pulse of ultraviolet radiation help make the electric discharge uniform and arc-free. Used in conjunction with rare gas halide and most molecular lasers.

Uncontrolled Area—An area to which access is not controlled for the purpose of protecting individuals from exposure to any laser radiation, and any area used as residential quarters.

Ventricular Fibrillation—A condition where the rhythmic pumping action of the heart stops and blood circulation ceases.

Visible Radiation (Light)—Electromagnetic radiation that can be sensed by the human eye. Generally used to describe wavelengths between 400 nm and 700 nm.

Visual Acuity—Keeness, sharpness or small-object-resolving capability of sight.

Vitreous Humor—A gelatinous substance that fills the portion of the eyeball between the retina and the lens.

Warning Logotype—A logotype as illustrated in the Laser Products Performance Standards.

Watt (W)—The unit of power or radiant flux; one joule per second.

Watt/cm^2 ($W \cdot cm^{-2}$)—A term used as a unit of incident power density or irradiance to express the amount of power per unit area of absorbing surface, or per unit area of a CW laser beam.

Wavelength (λ)—The distance between two points in a periodic wave which have the same phase.

Wavelength Selector—Optical device that selects a desired wavelength as the output of a multiline laser.

Yag—Yttrium aluminum garnet, a crystal that is doped with an active medium substance such as neodymium.

Yalo—Yttrium aluminate, a crystal that is doped with an active laser medium substance such as neodymium.

YLF—Yttrium lithium fluoride, a crystal that is doped with an active laser ionic substance such as holmium.

APPENDIX *B*

New York State Industrial Code Rule 50, "Lasers"

INDUSTRIAL
CODE RULE
50

LASERS

Part 50 of Title 12 of the Official Compilation of Codes,
Rules and Regulations of the State of New York
(Cited as 12 NYCRR 50)

Effective August 1, 1972

State of New York
Department of Labor
BOARD OF STANDARDS AND APPEALS

STATE OF NEW YORK

HUGH L. CAREY
Governor

DEPARTMENT OF LABOR

LOUIS L. LEVINE
Industrial Commissioner

BOARD OF STANDARDS AND APPEALS
Harry R. Mason, Chairman
B. Franklin Spencer, Member
Richard H. Bolton, Member

3-75 (1000)

STATE OF NEW YORK   }
DEPARTMENT OF LABOR   } ss.:
BOARD OF STANDARDS AND APPEALS}

In pursuance of the authority vested in me by subdivision 1 of section 102 of the Executive Law, I, HARRY R. MASON, Chairman of the New York State Board of Standards and Appeals, DO HEREBY CERTIFY that the copy of Industrial Code Part (Rule No.) 50 (12 NYCRR) relating to Lasers, hereto attached, is a correct transcript of the original of said rule duly adopted by said Board on the 13th day of June, 1972, prescribed to take effect on the 1st day of August, 1972, pursuant to the authority of sections 27-a, 28, 29 and 200 of the Labor Law, and duly filed in the office of the Secretary of State of the State of New York on the 9th day of July, 1972.

Given under my hand and the seal of office of the Department of Labor at the City of Albany, New York on the 12th day of July, 1972.

HARRY R. MASON
Chairman, New York State
Board of Standards and Appeals

[iii]

EXTRACTS FROM THE LABOR LAW

Section 27-a. General powers and duties of the board of standards and appeals. The board of standards and appeals shall have power, subject to the provisions of section twenty-nine of this chapter, to make, amend and repeal rules for carrying into effect the provisions of this chapter, applying such provisions to specific conditions and prescribing means, methods and practices to effectuate such provisions.

* * * *

Section 213. Violations of provisions of labor law; the industrial code; the rules, regulations or orders of the industrial commissioner and the board of standards and appeals. Any person who violates or does not comply with any provision of the labor law, any provision of the industrial code, any rule, regulation or lawful order of the industrial commissioner or the board of standards and appeals, and the officers of any corporation who knowingly permit the corporation to violate such provisions, are guilty of a misdemeanor....

* * * * *

EXPLANATION OF NUMBERING SYSTEM

In accordance with the uniform numbering system established by the Secretary of State for the *Official Compilation of Codes, Rules and Regulations of the State of New York* the following arrangement has been used in the text of Industrial Code Part (Rule No.) 50 (12 NYCRR 50) contained herein.

Division of Numbering System	How Numbered	Example
Title	Arabic numeral	Title 12 (Department of Labor)
Chapter	Capital roman numeral	Chapter I (Board of Standards and Appeals)
Subchapter	Capital letter	Subchapter A (The Industrial Code)
Part	Arabic numeral	Part 50 (Industrial Code Part (Rule No.) 50)
Section	Arabic numeral separated from Part number by decimal	50.1, 50.2 ...
Subdivision	Small letter in parentheses	(a), (b) ...
Paragraph	Arabic numeral in parentheses	(1), (2) ...
Subparagraph	Small roman numeral in parentheses	(i), (ii) ...
Clause	Small italic letter in parentheses	(*a*), (*b*) ...

References in or in connection with the following text to "Part," "section", "subdivision", "paragraph", "subparagraph" or "clause" are to be understood in accordance with the system outlined above.

[iv]

INDUSTRIAL CODE PART (RULE NO.) 50 (12 NYCRR 50)

CONTENTS

v

The Industrial Code of the State of New York
Part (Rule No.) 50
relating to

LASERS

(Statutory Authority: Labor Law §§ 27-a, 28, 29, 200)

Title 12 of the New York State Official Compilation of Codes,
Rules and Regulations, Chapter I, Subchapter A, Part 50

Section 50.1 Finding of fact. The board finds that certain industries, trades, occupations and processes involving the use or presence of lasers involve elements of potential danger to the lives, health and safety of persons employed therein. The board finds, therefore, that special regulations are necessary for the protection of such persons in that lasers emit infrared, visible or ultraviolet coherent light which may have the property of producing deleterious effects upon the human body and within the human eye and in that other hazards associated with lasers may be present.

50.2 Application. (a) *General.* Except as herein provided, this Part (rule) shall apply throughout the State of New York to every person subject to the jurisdiction of the Labor Law who, in any industry, trade, occupation or process, transfers, receives, possesses or uses any laser while such laser is free from and not subject to the regulatory powers and jurisdiction of the New York State Department of Health or the New York City Department of Health. This Part (rule) also applies to every person subject to the jurisdiction of the Labor Law who, in any industry, trade, occupation or process, engages in the installation, testing or servicing of any such laser or laser equipment that may result in the exposure of such person to laser radiation and other hazards associated with lasers.

(b) *Human use.* Nothing in this Part (rule) shall be construed as limiting the use of lasers in the healing of humans when done by or under the supervision of an individual licensed to practice medicine in the State of New York.

(c) *Animal use.* Nothing in this Part (rule) shall be construed as limiting the use of lasers in the healing of animals when done by or under the supervision of an individual licensed to practice veterinary medicine in the State of New York.

(d) *Federal statute.* Nothing in this Part (rule) affects requirements promulgated pursuant to any Federal statute and applicable in the State of New York.

50.3 Specific exemptions. The transfer, receipt, possession or use of lasers under the following conditions shall be exempt from the requirements of this Part (rule):

1

(a) *Exemption No. 1.* Lasers during the time period of their storage, shipment or sale when such equipment is not readily capable of emitting laser radiation. However, the laser labeling requirements of this Part (rule) shall apply during such time periods.

(b) *Exemption No. 2.* Lasers which by reason of their design and construction cannot emit radiation that exceeds 1×10^{-7} joules/cm^2 or 1×10^{-5} watts/cm^2 when measured 10 centimeters from the exterior surfaces of such lasers. This exemption does not apply to the testing or servicing of such lasers during their production or repair.

50.4 Definitions. As used herein or in connection with this Part (rule), the following terms mean:

(a) *Angstrom (Å).* A unit of length used mainly in expressing length of electromagnetic waves. An angstrom is equal to 10^{-10} meter, 10^{-8} centimeter or 10^{-4} micron.

(b) *Approved.* In respect to a device or material: in compliance with a subsisting resolution of approval adopted by the board; in respect to action by the board: made the subject of a resolution of approval.

(c) *Beam divergence.* The full angle of laser beam spread usually measured at the half power points in radians or milliradians.

(d) *Board.* The Board of Standards and Appeals of the State of New York.

(e) *Certificate of competence.* A document issued to the operator of a mobile laser by the commissioner in accordance with the provisions of this Part (rule).

(f) *Certified mobile laser operator.* Any individual holding a valid certificate of competence issued by the commissioner in accordance with the provisions of this Part (rule).

(g) *Closed installation.* Any installation where lasers are used and the entry to which is controlled while any laser therein is being operated.

(h) *Commissioner.* The Industrial Commissioner of the State of New York or his duly authorized representative.

(i) *C.W. laser.* A continuous wave laser.

(j) *Designated individual.* An individual selected and directed by a laser safety officer to supervise the operation of a laser.

(k) *Employee.* An individual employed; one who works for wages or salary in the service of another.

(l) *Energy density.* The energy per unit area expressed in units of Joules per square centimenter (J/cm^2).

(m) *Gas laser.* A type of laser in which the lasing action occurs in a gas medium, such as a helium-neon mixture.

2

(n) *High-intensity laser*. A laser with an output energy or power density that exceeds the values listed in Table 3 of this Part (rule).

(o) *Individual*. Any human being.

(p) *Irradiance*. An expression for incident power density.

(q) *Joule (J)*. A unit of energy equal to 10^7 erg or one watt-second.

(r) *Joule per square centimeter (J/cm²)*. The energy per unit area and a unit of energy density and radiant exposure.

(s) *Laser*. An acronym for light amplification by stimulated emission of radiation, also referred to as an optical maser.

(t) *Laser installation*. Any location where, for a period of more than 30 days, one or more lasers are used or operated. The confines of a laser installation shall be designated by the owner of such installation. An entire building or other structure, a part thereof or a plant may be designated as a laser installation. For the purposes of this Part (rule), a construction site shall not be considered a laser installation.

(u) *Laser radiation area*. Any area containing one or more high- or low-intensity lasers and the access to which is controlled for the purpose of protecting individuals from exposure to laser radiation.

(v) *Laser safety officer*. An individual, designated at a particular laser installation or for a particular mobile laser, who is qualified by training and experience in the occupational and public health aspects of lasers to evaluate the radiation hazards of such laser installation or mobile laser and who is qualified to establish and administer a laser radiation protection program for such laser installation or mobile laser.

(w) *Low-intensity laser*. A laser with an output energy or power density at any point along the path of the laser beam to which the eye may be exposed which is less than or equal to the values listed in Table 3 of this Part (rule).

(x) *Maser*. An acronym for microwave amplification by the stimulated emission of radiation. A maser emits microwave radiation instead of light.

(y) *Maximum permissible energy density for corneal and skin exposure*. The energy density of laser radiation from a pulsed laser which, in accordance with present medical knowledge, is not expected to cause detectable corneal or other bodily injury to an individual at any time during his lifetime. The maximum permissible levels are listed in Tables 1 and 2 of this Part (rule).

(z) *Maximum permissible power density for corneal and skin exposure*. The power density of laser radiation from a C.W. or pulsed laser which, in accordance with present medical knowledge, is not expected to cause detectable corneal or other bodily injury to an individual at any time during his lifetime. The maximum permissible levels are listed in Tables 1 and 2 of this Part (rule).

3

(aa) *Mobile laser*. A laser which is used or operated outside a laser installation.

(bb) *Nanometer (nm)*. A unit of length equal to 10^{-9} meter, 10^{-7} centimeter, 10^{-3} micron or 10 Angstroms. The nanometer is gradually replacing the Angstrom as the unit of measurement of electromagnetic wave lengths.

(cc) *Optical density (O.D.)*. A number equal to the common logarithm (base 10) of the attenuation afforded by a filter.

(dd) *Optically pumped lasers*. Types of lasers that derive energy from high-intensity light sources, such as xenon flash lamps.

(ee) *Output energy*. The energy emitted by a laser. This term is used to evaluate pulsed lasers.

(ff) *Output power*. Laser output defined by energy per unit time. This term is used to evaluate continuous wave lasers.

(gg) *Owner*. Any person conducting the business or activities carried on within a laser installation or with a mobile laser and having by law the administrative control of a laser, whether as owner, lessee, contractor or otherwise.

(hh) *Person*. Any of the following: an individual; a corporation; a partnership; or a firm.

(ii) *Power density*. The power per unit area usually expressed in watts per square centimeter (w/cm^2).

(jj) *Pulsed laser*. A laser that delivers its energy in pulses of short duration.

(kk) *Pulse length*. The duration of a pulsed laser flash measured in units of milliseconds (msec) which equals 10^{-3} second, microseconds (μsec) which equals 10^{-6} second, nanoseconds (nsec) which equals 10^{-9} second or picoseconds (psec) which equals 10^{-12} second.

(ll) *Pulsed re-occurrence frequency (P.R.F.)*. The frequency of occurrence of laser pulses in units of pulses per second. Pulsed re-occurrence frequency is also referred to as pulsed repetition frequency or rate.

(mm) *Q-switching Q-spoiling*. A technique used to obtain extremely high peak powers for very short durations from pulsed lasers.

(nn) *Radiant exposure*. An expression for incident energy density.

(oo) *Repetitive pulse laser*. A pulsed laser with re-occurring pulsed output.

(pp) *Research and development*. (1) Theoretical analysis, exploration or experimentation ; or

(2) The extension of investigative findings and theories of a scientific or technical nature into practical application for experimental, demonstrative and specialized purposes including the experimental or limited production and testing of models, devices, equipment, materials and processes involving the use of lasers.

4

(qq) *Semi-conductor or junction laser.* A type of laser in which the lasing action occurs in a semi-conductor, such as gallium arsenide. Such semi-conductors are sometimes cooled to cryogenic temperatures for more efficient operation.

(rr) *Shall.* The word "shall" is always mandatory.

(ss) *Shield.* An enclosure for a laser, a laser beam and/or a target. Typical laser shielding materials are rigid plastics, metals, opaque ceramics and light-impervious cloth. The term applies to a radiation shield, an explosion shield or to a combination radiation and explosion shield.

(tt) *Specular or regular reflection.* Mirror-like reflection.

(uu) *Survey.* An evaluation of the laser radiation hazards incident to the production, use, disposal, servicing or presence of any laser including, if appropriate, a physical evaluation of the laser installation or mobile laser and the measurement of laser output and reflected laser radiation.

(vv) *Trainee.* An individual, at least 18 years of age, who is being trained in the proper and safe use and operation of lasers.

(ww) *Uncontrolled area.* An area the access to which is not controlled for the purpose of protecting individuals from exposure to any laser radiation and any area used as residential quarters.

(xx) *Watt (w).* A unit of power.

(yy) *Watt per square centimeter (w/cm²).* The power per unit area and a unit of power density and irradiance.

50.5 Responsibility of employers. Every employer or owner shall effect compliance with the provisions of this Part (rule) relating to laser radiation and other associated hazards. No employer or owner shall suffer or permit any individual to be exposed to laser radiation above the maximum permissible levels specified in Tables 1 and 2 of this Part (rule) or to other associated hazards without such individual wearing the personal protective equipment required by this Part (rule).

50.6 Responsibility of employees. Every employee shall properly use the protective equipment provided for his protection and shall comply with all the provisions of this Part (rule) relating to his personal conduct.

50.7 Approval of low-intensity lasers required. (a) After January 1, 1973, any low-intensity laser used or operated in New York State, except for any such laser used exclusively in research and development, shall be approved.

(b) Approved low-intensity lasers are exempt from the registration requirement of section 50.8 of this Part (rule) but such lasers and their use shall be in compliance with other applicable provisions of this Part

5

(rule). Low-intensity lasers used exclusively in research and development are not exempt from the registration requirement of section 50.8 of this Part (rule) and are not exempt from the other applicable provisions of this Part (rule).

50.8 Registration of laser installations and mobile lasers. All laser installations and mobile lasers, including low-intensity lasers used exclusively in research and development, shall be registered with the commissioner at such times and on such forms as prescribed by him. After August 1, 1972, such registration shall be made prior to receipt or assembly of any laser. Laser installations and mobile lasers in use on or before August 1, 1972 shall be registered within 90 days after August 1, 1972.

> *Exception:* Laser installations and mobile lasers utilizing approved low-intensity lasers exclusively are exempt from this registration requirement.

50.9 Certificate of competence required by operators of mobile lasers.
(a) *General.* No individual shall operate, nor shall any person or employer designate any individual to operate, any mobile laser unless such individual holds a valid certificate of competence which has been issued by the commissioner in accordance with the provisions of this Part (rule).

> *Exceptions:* (1) Professional engineers and land surveyors licensed to practice in the State of New York.
> (2) Operators of mobile lasers used exclusively in research and development.

(b) *Categories of certificates of competence.* There shall be two categories of certificates of competence issued to operators of mobile lasers, as follows:

(1) *Class A certificates.* The holder of a Class A certificate of competence may operate any low-intensity mobile laser.

(2) *Class B certificates.* The holder of a Class B certificate of competence may operate any high-intensity or low-intensity mobile laser.

(c) *Application forms and photographs.* An application for a certificate of competence or for a renewal thereof shall be made on forms provided by the commissioner. Along with such application forms, every applicant shall submit photographs of himself in such numbers and sizes as the commissioner shall prescribe. Such photographs shall have been taken within 30 days of such submission.

(d) *Physical condition.* No person suffering from an uncontrolled physical handicap or illness, such as epilepsy, heart disease, or from an uncorrected defect in vision or hearing, which might diminish his competence in the operation of laser equipment, shall be certified by the commissioner.

6

(e) *Age and experience required.* (1) Every applicant for a certificate of competence shall be at least 18 years of age.

(2) Every applicant for a certificate of competence shall have at least one year of practical experience in the operation of a laser. Such required experience may be obtained by a trainee who is designated by and working under the direct personal supervision of an individual permitted by this Part (rule) to operate such mobile laser equipment. Such experience shall include a knowledge of laser safety precautions and shall be acceptable to the commissioner. The commissioner may waive the one-year experience requirement if the applicant submits evidence of his successful completion of a laser operator training course acceptable to the commissioner.

(f) *Examining board.* (1) The commissioner shall appoint an examining board which shall consist of at least three members. At least one member of such board shall hold a valid Class B certificate of competence to operate a mobile laser.

(2) The members of such examining board shall serve at the pleasure of the commissioner and their duties shall include:

(i) The examination of applicants and their qualifications and the making of recommendations to the commissioner with respect to the experience and competence of such applicants.

(ii) The holding of hearings regarding appeals following denials of certificates.

(iii) The holding of hearings prior to determinations of the commissioner to suspend or revoke certificates or to refuse to issue renewals of certificates.

(iv) The reporting of findings and recommendations to the commissioner with respect to such hearings.

(3) The commissioner may designate panels to conduct such examinations and hearings on behalf of the examining board. Each such designated panel shall consist of at least three examining board members.

(4) The acts and proceedings of such examining board and of such panels shall be in accordance with regulations issued by the commissioner.

(g) *General examination required.* Each applicant for a certificate of competence shall, and each applicant for a renewal of such certificate may, be required by the commissioner to take an appropriate general examination.

Exceptions: The commissioner may issue a certificate of competence or a renewal thereof without examination of any kind to any applicant who:

(1) has successfully completed a four-year course of study at a college or university which included a

7

study of lasers and laser safety precautions and such course is acceptable to the commissioner; or

(2) has filed an application for a certificate of competence prior to February 1, 1973 and has submitted along with such application evidence of adequate experience with lasers and with laser safety precautions and who,. in the judgment of the commissioner, meets all other qualifications required for certification as a competent mobile laser operator.

(h) *Contents of a certificate of competence.* Each certificate of competence issued by the commissioner shall include the name and address of the certified mobile laser operator, a brief physical description and photograph of such operator for identification purposes.

(i) *Term of certificate of competence.* Each such certificate or renewal thereof shall be valid for three years from the date of issuance, unless such term is extended, suspended or revoked by the commissioner.

(j) *Carrying certificate.* Each certified mobile laser operator shall carry his certificate of competence on his person whenever he is working with or operating any mobile laser. Failure to produce such certificate upon request by the commissioner shall be considered presumptive evidence that the operator is not certified.

(k) *Certificate renewal.* An application for the renewal of a certificate of competence shall be filed with the commissioner not more than six months nor less than three months prior to the expiration date of the certificate sought to be renewed, except that the commissioner may alter such time limitation to prevent undue hardship to any certified mobile laser operator.

(l) *Suspension, revocation, refusal to renew and denial of certificates of competence; hearings.* (1) The commissioner may, upon notice to the interested parties and after a hearing before the examining board, suspend or revoke a certificate of competence upon finding that the certified mobile laser operator is not an individual of proper competence, judgment or ability in respect to mobile laser operation or for other good cause shown.

(2) Prior to a determination by the commissioner not to renew a certificate of competence, the commissioner shall require that a hearing be held by the examining board concerning such renewal application and that a notice of such hearing be sent to every interested party.

(3) Any applicant whose application for a certificate of competence has been denied by the commissioner may, upon his written request to the commissioner, have a hearing before the examining board. Such written request shall be made within 30 days after receipt of the notice of denial.

(4) Where a hearing has been held in connection with the suspension, revocation, refusal to renew or denial of a certificate of com-

8

petence, the examining board shall make its recommendation to the commissioner within three days after the conclusion of such hearing. A written notice of the commissioner's decision, containing the reasons therefor, shall be forwarded promptly to the applicant or certified mobile laser operator, as the case may be, as well as to any interested party who appeared at the hearing.

50.10 Maximum permissible exposure limits. (a) No person shall operate any laser so as to cause the cornea of the eye of any individual to be exposed to laser radiation beyond the limits listed in Table 1 of this Part (rule).

(b) No person shall operate any laser so as to cause the skin of any individual to be exposed to laser radiation beyond the limits listed in Table 2 of this Part (rule).

(c) All high-intensity lasers shall be placed in shielded and inter-locked enclosures or operated only in laser radiation areas.

Exceptions: (1) No such high-intensity laser need be enclosed, interlocked or operated in a laser radiation area if the commissioner is notified at least 48 hours before such planned operation on a form prescribed by him. A laser safety officer or designated individual shall be continuously present when such laser is operating.

(2) When repairing or servicing a high-intensity laser, the commissioner need not be so notified. However, a laser safety officer or designated individual shall be present during any period of operation without shielding or without interlocks in service. Such laser safety officer or designated individual shall insure that adequate precautions are taken to protect all individuals against exposure to hazardous laser radiation.

(d) The laser radiation exposure values produced by scanned laser beams shall be determined when such beams are operating in the scanning modes, providing the devices containing the beams are designed and constructed so as to prohibit the beams from operating in non-scanning modes.

50.11 Laser safety officer. The owner of any laser installation or mobile laser shall designate a laser safety officer who shall establish and administer a laser radiation safety program which is in compliance with this Part (rule). The laser safety officer for a high-intensity mobile laser may be the certified operator of such laser. The laser safety officer for one or more low-intensity lasers at a given site or location may be a certified mobile laser operator (Class A). When a laser safety officer does not personally supervise the safety aspects of laser operation, he may designate another individual, who has been sufficiently instructed and trained

9

by the laser safety officer in appropriate safety techniques, to personally supervise the operation of a laser. The laser safety officer or any individual he designates to supervise the safety aspects of laser operation may also operate laser equipment subject to the other applicable provisions of this Part (rule).

50.12 Personal protection. (a) *General.* Each person who possesses a laser shall instruct and advise every individual employed in or lawfully frequenting a laser radiation area in regard to the following:

(1) The presence of a laser in such area.

(2) The potential hazards associated with the use of the laser and the precautions and procedures necessary to minimize exposure to radiation from the laser.

(3) The applicable provisions of this Part (rule) for the protection of such individual from exposure to radiation from the laser.

(b) *Personal protective devices.* (1) *Approved safety eyewear.* Approved safety eyewear, as required by this Part (rule), shall be provided by the owner or employer and shall be used by the individuals working with or operating any unshielded laser and by other individuals lawfully frequenting the laser radiation area who may be exposed to laser radiation under circumstances where the conditions of laser use can lead to accidental exposure to radiation above the maximum permissible exposure limits listed in this Part (rule). Such safety eyewear shall meet the following specifications:

(i) The optical density of such approved safety eyewear shall reduce the external laser radiation to the cornea of the eye to safe levels as listed in Table 4 of this Part (rule).

(ii) Such safety eyewear shall be designed and tested to insure that the eyewear retains its protective properties during use.

(iii) Such safety eyewear shall be legibly labeled with the optical density of the lens and the wave length at which it was measured.

(iv) Any individual requiring prescription lenses in the normal performance of his work shall be provided with approved prescription eye protection or with approved safety eyewear designed to fit over his regular prescription lenses.

(2) *Other personal protective devices.* Protective gloves, clothing and shields shall be provided by the owner or employer and used by individuals as determined by the laser safety officer.

50.13 Special precautions. (a) No individual shall look directly into a primary laser beam or directly at the specular reflections of such a beam when the intensities of such beam or of such reflections are greater than those listed in Table 1 of this Part (rule), unless such individual is wearing approved eye protection in compliance with the provisions of this Part (rule).

10

(b) Every laser beam shall be terminated whenever necessary by a material that is non-reflecting and fire-retardant. In cases of intentional beam interaction with targets, precautions shall be taken to prevent fires arising from flying particles. In cases of unavoidable reflections, precautions shall be taken to prevent the scatter of laser radiation in excess of the limits listed in this Part (rule) into uncontrolled areas.

(c) Under circumstances where the conditions of laser operation can lead to accidental exposure to laser radiation above the maximum permissible exposure limits as specified in this Part (rule), an area shall be cleared of all individuals for a reasonable distance on all sides of the anticipated path of the laser beam, except for those individuals protected in compliance with the provisions of this Part (rule).

(d) Special care shall be taken to assure that approved safety eyewear is matched to the specific wave length of the laser device with which it is to be used and that such eyewear provides the required attenuation.

(e) Any approved safety eyewear which has been exposed to high-intensity laser radiation shall not be re-used until it has been re-evaluated for shattering and effectiveness of its attenuation.

(f) Laser beam paths shall be cleared of all material or objects that may cause the scatter of laser radiation at levels in excess of the limits listed in this Part (rule) into uncontrolled areas.

(g) All unshielded lasers which are capable of emitting radiation above the maximum permissible exposure limits as listed in this Part (rule) shall be capped or otherwise effectively terminated when not in operation.

(h) Interlocks used on enclosures for high-intensity lasers shall not be used to actuate such lasers.

(i) Any closed laser installation with one or more access openings large enough for entry by individuals shall have at least one such opening provided with an exit door which can be manually opened with ease from inside the laser installation.

(j) General illumination in laser radiation areas shall be at least 30 lumens per square foot, except where conditions of laser operation require lower ambient illumination.

(k) Electronic firing systems for pulsed lasers shall be so designed that accidental discharges of stored charges are prevented.

(l) "Fail-safe" safety circuitry shall be used with lasers wherever necessary.

50.14 Designation of laser radiation area. (a) *Laser radiation symbol.* The laser radiation symbol, as shown below, shall be used to designate lasers and laser radiation areas. The symbol shall be in red and the background area shall be yellow or white.

11

(b) *Laser radiation area.* Except as below provided, each person who possesses a laser shall post conspicuously at the entrance to and inside of each laser radiation area a sign constructed of durable material bearing the laser radiation symbol and the words: "CAUTION" or "DANGER" and "LASER RADIATION AREA". Such words shall be in black letters at least one inch in height.

> *Exception:* No such area need be posted with any sign or equipped with any required control device solely because of the presence therein of any laser for any period of not more than eight hours, provided, however, that during such period the laser safety officer or a designated individual is constantly in attendance and such safety officer or designated individual shall take all precautions necessary to prevent any individual from receiving any exposure to laser radiation that exceeds the applicable limits permitted by this Part (rule) and from receiving any other injury from any other hazard associated with the operation of any laser. Such area shall be subject to the control of the owner of such laser.

50.15 Designation of lasers. (a) Every laser not otherwise exempt by this Part (rule) shall have a label so attached as to be clearly visible and such label shall contain the following information:

(1) The laser radiation symbol in red.

12

(2) The words "CAUTION" or "DANGER" and "LASER" in black letters.

(3) The wave length or lengths of the emitted laser radiation.

(4) The maximum output power or energy density or densities of the emitted laser radiation.

(5) The divergence of the emitted laser radiation in its lowest order transverse mode.

(b) Such required information shall be legibly stamped, etched or otherwise permanently marked on a yellow or white label with letters and numbers not less than three millimeters (one-eighth inch) in height where practicable.

> *Exceptions:* (1) If such labeling of individual lasers is impractical because of laser size or location, such lasers shall be appropriately labeled in an alternative manner.
>
> (2) The label on any laser used exclusively in research and development need not specify numerically the wave length, maximum output power or energy density or the divergence where such parameters may be unknown or difficult to obtain. However, such labels shall contain the laser radiation symbol and the other wording required by paragraphs (1) and (2) of subdivision (a) of this section.

50.16 Surveys, instrumentation and inventories. (a) *Surveys.* Each person who possesses a registered laser shall make, or shall cause to be made, the applicable surveys required by this section and such additional surveys as may be necessary for such person to comply with the other provisions of this Part (rule) or as the commissioner may direct. Such surveys shall include:

(1) A survey of the output power or energy density of each registered mobile laser. Such determination shall be made by measurement or by calculation based on measurements made prior to initial use. If the laser beam is passed through an optical system, such determination shall be made at the point of maximum power or energy density along the path of the laser beam which is exposed to view.

(2) A survey shall be made of all laser protective devices, including safety interlocks, prior to initial use to make certain that such devices are in good working order and have been properly installed.

> *Note:* For lasers installed or operated prior to August 1, 1972, the surveys cited in paragraphs (1) and (2) of subdivision (a) of this section which are required to be performed prior to initial laser operation, shall be performed prior to the registration of such lasers.

(3) An inspection shall be made at least once every six months of all safety eyewear in use to make certain that the optical density and

13

type of filter is appropriate for the laser in use and to determine visually that the eyewear has no optical defects. All safety eyewear found to be deteriorated or defective in any way shall be immediately withdrawn from use and repaired if possible or otherwise discarded.

(b) *Instrumentation.* All the necessary measurements performed in compliance with this Part (rule) shall be made with instrumentation designed and calibrated for use with the laser that is under surveillance.

(c) *Inventories.* An inventory shall be made of all lasers in any laser installation and all mobile lasers at least once a year. Such inventories shall be kept available for examination by the commissioner.

50.17 Safeguarding and disposing of lasers. (a) *Safeguarding.* Each laser shall be safeguarded against unauthorized use.

(b) *Disposing.* No person shall dispose of a high-intensity laser except by making such laser permanently inoperative or by transferring such laser to another person. The recipient of any such transferred laser, if subject to the provisions of the Labor Law, shall obtain the required registration in accordance with subdivision (a), section 50.8 of this Part (rule) and shall display the proof of such registration to the person disposing of such high-intensity laser.

50.18 Associated hazards. In addition to the protection against laser radiation, protection shall also be provided against any associated hazards which may be present during the operation of any laser. Such associated hazards include but are not limited to the following:

(a) *Air contaminants.* Vaporized target materials and toxic gases, vapors and fumes which may be associated with any laser may be present in a laser radiation area. Such dangerous air contaminants shall be removed or controlled in accordance with the provisions of Industrial Code Part (Rule No.) 12 relating to "Control of Air Contaminants" and Industrial Code Part (Rule No.) 18 relating to "Exhaust Systems". Adequate ventilation shall be maintained in each laser installation.

(b) *Ultra-violet radiation.* Ultra-violet radiation levels in excess of 0.5 microwatt/cm^2 for seven hours or 0.1 microwatt/cm^2 for continuous exposure shall be avoided. Because quartz transmits ultra-violet radiation efficiently, particular care shall be taken to protect individuals from such radiation when quartz tubes are used in the laser system.

(c) *Electrical hazards.* Any laser and its associated electrical system may present electrical hazards. Every laser, including all associated electrical equipment, shall be so designed, constructed, installed and maintained as to minimize the possibility of electrical hazards.

(d) *Cryogenic coolants.* The storage, handling and use of cryogenic coolants associated with lasers shall be such so as to minimize any hazards which such coolants may present.

14

(e) *Fire hazards.* Every laser and every laser installation shall be designed, constructed, installed, operated and maintained so as to eliminate or reduce any fire hazard. In addition, every laser radiation area shall be maintained clear of any unnecessary materials in order to minimize any fire hazard.

(f) *Explosion hazards.* Lasers and associated equipment may present explosion hazards. Such equipment shall be protected with explosion shields to prevent injury to individuals from such hazards.

(g) *Ionizing radiation.* A laser system may produce ionizing radiation in which case such system shall be in compliance with the provisions of Industrial Code Part (Rule No.) 38 relating to "Radiation Protection" in respect to protection against ionizing radiation.

50.19 Records. Each person who possesses a registered laser shall maintain accurate and complete written records which shall be kept available for examination by the commissioner. Such records shall be kept of the following:

(a) The results of each required initial laser output check and system interlock check.

(b) Each transfer, receipt and disposal of any registered laser.

(c) Annual inventories, safety eyewear checks and equipment measurements obtained in accordance with the required surveys and instrumentation checks listed in section 50.16 of this Part (rule).

50.20 Reports. (a) *Reports to the commissioner.* Each person who possesses a laser shall report to the commissioner immediately upon the occurrence of any of the following:

(1) Any theft or loss of an intact laser.

(2) Any injury to an individual resulting from the operation of a laser or of any associated equipment. In addition, a written report of such injury shall be forwarded to the commissioner within seven days of such injury.

(b) *Reports to physicians.* In case of a required reportable exposure to laser radiation, all pertinent data relating to such exposure shall be made available to any physician who has been authorized by the exposed individual to receive such data.

50.21 Inspection and tests. (a) Each person who possesses a laser shall permit the commissioner to inspect, except as noted in subdivision (c) below, the following at all reasonable times:

(1) The laser, the laser radiation area and the laser installation or premises where such laser is located, possessed, stored or used.

(2) Each record required to be maintained by the provisions of this Part (rule).

15

(b) Each person who possesses a laser shall permit the commissioner to conduct such tests as he may deem necessary at any reasonable time except as noted in subdivision (c) below. Such tests shall utilize the advice and assistance of the laser safety officer where appropriate.

(c) If any inspection or test conducted in accordance with subdivisions (a) or (b) above may compromise national security, the commissioner shall, in lieu of such inspection or test, accept a statement from the owner and suitable reports from the laser safety officer concerning the proper operation of the laser and the safety precautions in force.

50.22 Severability. If any provision of this Part (rule) or the application thereof to any person or circumstance is held invalid, such invalidity shall not affect other provisions or applications of this Part (rule) which can be given effect without the invalid provision or application and to this end the provisions of this Part (rule) are declared to be severable.

50.23 Tables. The tables hereto annexed and designated: "Table 1—Maximum Permissible Corneal Exposure for Direct Illumination or Specular Reflection of Laser Radiation"; "Table 2—Maximum Permissible Skin Exposure"; "Table 3—Maximum Permissible Output Densities for Non-Enclosed Laser in an Uncontrolled Area (Limits for Low-Intensity Lasers)"; and "Table 4—Attenuation of Laser Safety Eyewear" are hereby made part of this Part (rule).

TABLE 1

MAXIMUM PERMISSIBLE CORNEAL EXPOSURE FOR DIRECT ILLUMINATION OR SPECULAR REFLECTION OF LASER RADIATION *

Q-switched 1 nanosecond to 1 microsecond pulse [10^{-9}–10^{-6} sec.] (joules/cm^2)	*Non-Q-switched 1 microsecond to 0.1 of a second pulse [10^{-6}–10^{-1} sec.] (joules/cm^2)*	*Continuous wave or pulse width greater than 0.1 of a second duration [Greater than 10^{-1} sec.] (watts/cm^2)*
1.0 x 10^{-7} ** 1.0 x 10^{-3} ***	1.0 x 10^{-6} ** 1.0 x 10^{-2} ***	1.0 x 10^{-5} ** 1.0 x 10^{-2} ***

* The values in this table assume a diffraction limited and zero order transverse mode beam. In the case of a higher order transverse mode in a gas laser where the intrinsic beam divergence exceeds one milliradian the values may be increased on the basis of experimental evidence. Because of the lack of data, no explicit value is listed for the maximum permissible corneal exposure for laser radiation in the wave length range below 400 nm. or for sub-nanosecond pulses; therefore, the cornea shall not be directly exposed to this radiation until MPCE values have been established for these conditions.

** For lasers operating in the wave length range 400–1400 nm.

*** For lasers operating in the wave length range above 1400 nm.

16

TABLE 2

MAXIMUM PERMISSIBLE SKIN EXPOSURE *

Energy Density, Pulsed [Q-switched, Non-Q-switched] (joules) (cm²)	Power Density, Continuous Wave (watts) (cm²)
0.1	1.0

* For lasers operating in the visible, near-infrared and infrared regions of the electromagnetic spectrum. No information is available on permissible skin exposures to laser radiation in the wave length range below 400 nm. Therefore, caution must be used to avoid such exposure until experimental data is available on permissible levels.

TABLE 3

MAXIMUM PERMISSIBLE OUTPUT POWER OR ENERGY DENSITIES FOR A NON-ENCLOSED LASER IN AN UNCONTROLLED AREA (LIMITS FOR LOW-INTENSITY LASERS)

Q-switched 1 nanosecond to 1 microsecond pulse [10^{-9}–10^{-6} sec.] (joules/cm²)	Non-Q-switched 1 microsecond to 0.1 of a second pulse [10^{-6}–10^{-1} sec.] (joules/cm²)	Continuous wave or pulse width greater than 0.1 of a second duration [Greater than 10^{-1} sec.] (watts/cm²)
1.0×10^{-1}	1.0	3.0

17

TABLE 4

ATTENUATION OF LASER SAFETY EYEWEAR *

O.D.	Attenua-tion (db) **	Attenua-tion Factor	Suggested Maximum Incident Density		
			Q-switched Max. Energy Density (J/cm^2)	Non-Q-switched Max. Energy Density (J/cm^2)	Continuous Wave Maximum Power Density (W/cm^2)
1	10	10	10^{-6}	10^{-5}	10^{-4}
2	20	10^2	10^{-5}	10^{-4}	10^{-3}
3	30	10^3	10^{-4}	10^{-3}	10^{-2}
4	40	10^4	10^{-3}	10^{-2}	10^{-1}
5	50	10^5	10^{-2}	10^{-1}	1
6	60	10^6	10^{-1}	1	10
7	70	10^7	1	10	—
8	80	10^8	10	10^2	—

* Because of limitations of safety eyewear material, no eyewear shall be exposed to more than 400 joules or 10 watts of incident laser energy or power, respectively.

** (db) = Decibel. A unit to express a beam intensity ratio. The decibel is equal to 10 times the logarithm of the beam intensity ratio expressed by the following equation:

$$n(db) = 10 \log_{10} \frac{(P_1)}{(P_2)}$$

where P_1 and P_2 designate the input and output power density or energy density, respectively, and "n" designates the number of decibels corresponding to their ratio.

18

APPENDIX *C*

Class Level of Accessible Laser Radiation

Class level determination of laser radiation is best explained using specific examples of radiation and sequentially invoking the requirements of Section 1040.10 (d) (i.e., Accessible Emission Limits) and other appropriate requirements of the Laser Products Performance Standards. Thus, Example C-1, which follows, is included in Appendix C.

Provision has been made in the immediate discussion for considering uncertainties as required by Section 1040.10 (e) (1) (i.e., Tests For Determination of Compliance—Tests for Certification) and for considering the aperture-stop test conditions as required by Sections 1040.10 (e) (3) and (4) (i.e., Tests for Determination of Compliance—Measurement Parameters and Measurement Parameters for Scanned Laser Radiation).

Example C-1. CW Radiation of Multiple Wavelength.[1] Consider a multi-wavelength, unscanned laser beam emitted continuously. All of the radiation is uniformly distributed within a circular area 80 mm in diameter (50.265 cm^2). The various wavelengths' measured radiant energy and levels, not including a probable error of measurement of ±15% expressed at a 3σ (95%) confidence level, are:

[1] Adapted from Reference 5 of Chapter 10.

Wavelength (nm)	Radiant Energy Level (Joules)	Radiant Exposure Level $(J \cdot cm^{-2})$
10,600	$6.20 \times 10^{-4} \ t$	$1.23 \times 10^{-5} \ t$
1,100	$1.60 \times 10^{-4} \ t$	$3.18 \times 10^{-6} \ t$
750	$1.30 \times 10^{-5} \ t$	$2.59 \times 10^{-7} \ t$

Step 1—General Considerations. The components of the composite laser beam fall within two wavelength ranges:

- The range of wavelengths $> 1,400$ nm but $\leq 13,000$ nm
- The range of wavelengths > 700 nm but $\leq 1,400$ nm

The class level of each wavelength component must be considered separately, as well as the class level of each wavelength range containing two or more wavelength components.

Step 2—Consider Wavelength Range Containing Only 10,600-nm Component

A) From Table 7-2A of Chapter 7: $k_1 = 1.0$ and $k_2 = 1.0$.
B) Emission Duration Ranges (Column 4 of Table C-1): Examination of the accessible emission limits for the Class I level (Table 7-1A of Chapter 7) for $\lambda = 10,600$ nm indicates that the emission duration ranges to be considered are:

$$1.0 \times 10^{-9} \ sec < t \leq 1.0 \times 10^{-7} \ sec$$

$$1.0 \times 10^{-7} \ sec < t \leq 10 \ sec$$

$$10 \ sec < t$$

C) Accessible Emission Limits (Column 5 of Table C-1): From Table 7-1A, for $k_1 = k_2 = 1$ and for the respective emission duration ranges, the accessible emission limits follow:

$$7.9 \times 10^{-5} \ J$$

$$4.4 \times 10^{-3} \ t^{1/4} \ J$$

$$7.9 \times 10^{-4} \ t \ J$$

D) Accessible Emission Levels (Column 6 of Table C-1): Obtain accessible emission levels that correspond to the highest relative accessible emission levels.

- This is done by selecting the highest emission duration for the particular emission duration range:

$$1.0 \times 10^{-7} \text{ sec}$$

$$10 \text{ sec}$$

$$10 \text{ sec} < t$$

Enter values in Column 8 of Table C-1 (Emission Duration—sec).

From given information: Radiant Energy Level = 6.20×10^{-4} t J (i.e., Accessible Emission Level)

for $t = 1.0 \times 10^{-7}$ sec: 6.20×10^{-4} (1.0×10^{-7}) J = 6.2×10^{-11} J

for $t = 10$ sec   : 6.20×10^{-4} (10) J = 6.2×10^{-3} J

for 10 sec $< t$: 6.20×10^{-4} t J

E) Relative Accessible Emission Levels $\left[\dfrac{E(\lambda, t)}{E(N, \lambda, t)_{\max}} \right]$ (Column 7 of Table

C-1): Obtain ratios for respective values of Items (C) and (D) above.

$$\frac{6.2 \times 10^{-11} \text{ J}}{7.9 \times 10^{-5} \text{ J}} = 7.8 \times 10^{-7}$$

$$\frac{6.2 \times 10^{-3} \text{ J}}{4.4 \times 10^{-3} \; t^{1/4} \text{ J}} = \frac{1.41}{(10)^{1/4}} = \frac{1.41}{1.78} = 0.79$$

$$\frac{6.2 \times 10^{-4} \; t \text{ J}}{7.9 \times 10^{-4} \; t \text{ J}} = 0.79$$

F) Estimated Error in Value Reported in Column 6 and Confidence Level (Column 11 of Table C-1)

$$\pm 15\% \text{ of } 6.2 \times 10^{-11} \text{ J}: \pm 0.93 \times 10^{-11} \text{ J } (3\sigma)$$

$$\pm 15\% \text{ of } 6.2 \times 10^{-3} \text{ J}: \pm 0.93 \times 10^{-3} \text{ J } (3\sigma)$$

$$\pm 15\% \text{ of } 6.2 \times 10^{-4} \; t \text{ J}: \pm 0.93 \times 10^{-4} \; t \text{ J } (3\sigma)$$

G) Relative Accessible Emission Levels for 15% Estimated Measurement Error

$$\frac{(6.2 \times 10^{-11} \text{ J}) + (0.93 \times 10^{-11} \text{ J})}{7.9 \times 10^{-5} \text{ J}} = 9.0 \times 10^{-7}$$

$$\frac{(6.2 \times 10^{-3} \text{ J}) + (0.93 \times 10^{-3} \text{ J})}{4.4 \times 10^{-3} \; t^{1/4} \text{ J}} = \frac{1.62}{1.78} = 0.91$$

$$\frac{(6.2 \times 10^{-4} \; t \text{ J}) + (0.93 \times 10^{-4} \; t \text{ J})}{7.9 \times 10^{-4} \; t \text{ J}} = 0.90$$

Table C-1. Accessible Laser Radiation Parameters.

1 Radiation Ident.	2 Class Level	3 Wavelength (nm)	4 Emission Duration Range (sec)	5 Accessible Emission Limit	6 Accessible Emission Level	7 $\dfrac{E(\lambda, t)}{E(N, \lambda, t)_{max}}$	8 Emission Duration (sec)	9 Aperture Stop Diam. (mm)	10 Meas. Instr. Ident.	11 Estimated Error In Value Reported In Column 6 And Confidence Level	12 Necessary For Function "Y" or "N"	13 Access Code "O" or "S"
Example 1	I	750	$1.0 \times 10^{-9} < t < 2.0 \times 10^{-5}$	2.5×10^{-7} J	2.60×10^{-10} J	1.04×10^{-3}	2.0×10^{-5}	80	–	$\pm 0.39 \times 10^{-10}$ J (3σ)	–	–
			$2.0 \times 10^{-5} < t \leqq 10$	$8.75 \times 10^{-4}\, t^{3/4}$ J	1.30×10^{-4} J	0.0264	10	80	–	$\pm 0.195 \times 10^{-4}$ J (3σ)	–	–
			$10 < t < 198$	4.88×10^{-3} J	2.57×10^{-3} J	0.527	198	80	–	$\pm 0.386 \times 10^{-3}$ J (3σ)	–	–
			$198 < t < 1.0 \times 10^4$	$2.46 \times 10^{-5}\, t$ J	$1.30 \times 10^{-5}\, t$ J	0.528	$198 < t < 1.0 \times 10^4$	80	–	$\pm 0.195 \times 10^{-5}\, t$ J (3σ)	–	–
			$t > 1.0 \times 10^4$	$2.46 \times 10^{-5}\, t$ J	$1.30 \times 10^{-5}\, t$ J	0.528	$1.0 \times 10^4 < t$	80	–	$\pm 0.195 \times 10^{-5}\, t$ J (3σ)	–	–
	I	1,100	$1.0 \times 10^{-9} < t < 2.0 \times 10^{-5}$	1.0×10^{-6} J	3.20×10^{-9} J	0.0032	2×10^{-5}	80	–	$\pm 0.48 \times 10^{-9}$ J (3σ)	–	–
			$2.0 \times 10^{-5} < t < 10$	$3.5 \times 10^{-3}\, t^{3/4}$	1.6×10^{-3} J	0.081	10	80	–	$\pm 0.24 \times 10^{-3}$ J (3σ)	–	–
			$10 < t < 100$	1.95×10^{-2} J	1.6×10^{-2} J	0.820	100	80	–	$\pm 0.24 \times 10^{-2}$ J (3σ)	–	–
			$100 < t < 1.0 \times 10^4$	$1.95 \times 10^{-4}\, t$ J	$1.6 \times 10^{-4}\, t$ J	0.820	$100 < t < 1.0 \times 10^4$	80	–	$\pm 0.24 \times 10^{-4}\, t$ J (3σ)	–	–
			$1.0 \times 10^4 < t$	$1.95 \times 10^{-4}\, t$ J	$1.6 \times 10^{-4}\, t$ J	0.820	$1.0 \times 10^4 < t$	80	–	$\pm 0.24 \times 10^{-4}\, t$ J (3σ)	–	–
	I	10,600	$1.0 \times 10^{-9} < t < 1.0 \times 10^{-7}$	7.9×10^{-5} J	6.2×10^{-11} J	7.8×10^{-7}	1.0×10^{-7}	80	–	$\pm 0.93 \times 10^{-11}$ J (3σ)	–	–
			$1.0 \times 10^{-7} < t < 10$	$4.4 \times 10^{-3}\, t^{1/4}$ J	6.2×10^{-3} J	0.79	10	80	–	$\pm 0.93 \times 10^{-3}\, t$ J (3σ)	–	–
			$10 < t$	$7.9 \times 10^{-4}\, t$ J	$6.2 \times 10^{-4}\, t$ J	0.79	$10 < t$	80	–	$\pm 0.93 \times 10^{-4}\, t$ J (3σ)	–	–
III b	Combined 750 nm and 1,100 nm		–	–	–	–	–	–	–	–	–	–

This shows that the 10,600-nm component is rather close to the Class I limit when the probable error is considered because the relative accessible emission level becomes 0.91.

H) Miscellaneous Columns of Table C-1

- Column 1 (Radiation Identification): Example 1
- Column 2 (Class Level): I
- Column 3 (Wavelength in nm): 10,600
- Column 9 (Aperture Stop Diameter in mm): 80
- Column 10 (Measurement Instrumentation Identification): Insert dash (–) as this is an example only and no instruments were actually used to measure the accessible laser radiation.
- Column 12 (Necessary for Function "Y" or "N"): Insert dash (–) as this is an example only, and it cannot be stated whether the level of radiation is necessary for the intended function of the laser product.
- Column 13 (Access Code "O" or "S"): Insert dash (–) as this is an example only, and it cannot be stated whether access to radiation is necessary during operation of the product, or if access to radiation is necessary only during service of the product.

Step 3—Consider Wavelength Range Containing Only 750-nm Component

A) Wavelength-Dependent Factors k_1 and k_2 (from Table 7-2A)

$$k_1 = 10^{[(\lambda - 700)/515]} \quad \text{for all values of } t$$

$$= 10^{[(750-700)/515]} = 10^{0.097}$$

$$= 1.25$$

$$k_2 = 1.0 \quad \text{for } t \leq \frac{10,100}{\lambda - 699} \leq \frac{10,100}{750 - 699} \leq 198 \text{ sec}$$

also

$$k_2 = \frac{t(\lambda - 699)}{10,000} \quad \text{for } \frac{10,100}{\lambda - 699} < t \leq 10^4$$

$$= \frac{t(750 - 699)}{10,100} \quad \frac{10,100}{750 - 699} < t \leq 10^4$$

$$= 0.00505 \, t \quad \text{for } 198 \text{ sec} < t \leq 10^4 \text{ sec}$$

and:

$$k_2 = \frac{\lambda - 699}{1.01} \quad \text{for } t > 10^4$$

$$= \frac{750 - 699}{1.01}$$

$$= 50.5 \quad \text{for } t > 10^4 \text{ sec}$$

In summary:

$$k_1 = 1.25 \qquad \text{for all values of } t$$
$$k_2 = 1.0 \qquad \text{for } t \leq 198 \text{ sec}$$
$$k_2 = 0.00505\,t \qquad \text{for } 198 \text{ sec} < t \leq 10^4 \text{ sec}$$
$$k_2 = 50.5 \qquad \text{for } t > 10^4 \text{ sec}$$

B) Emission Duration Ranges (Column 4 of Table C-1): Examination of the accessible emission limits for the Class I level (Table 7-1A) and the above k_1 and k_2 indicates the emission duration ranges to be considered for $\lambda = 750$ nm are:

$$1.0 \times 10^{-9} \text{ sec} < t \leq 2.0 \times 10^{-5} \text{ sec}$$

$$2.0 \times 10^{-5} \text{ sec} < t \leq 10 \text{ sec}$$

$$10 \text{ sec} < t \leq 198 \text{ sec}$$

$$198 \text{ sec} < t \leq 1.0 \times 10^4 \text{ sec}$$

$$1 \times 10^4 \text{ sec} < t$$

C) Accessible Emission Limits (Column 5 of Table C-1): From Table 7-1A, using appropriate k_1 and k_2 the accessible emission limits follow for the emission duration ranges shown:

- For 1.0×10^{-9} sec $< t \leq 2.0 \times 10^{-5}$ sec:

$$2.0 \times 10^{-7} \, k_1 k_2 \text{ J}$$

$$= 2.0 \times 10^{-7} \, (1.25)\,(1) \text{ J}$$

$$= 2.50 \times 10^{-7} \text{ J}$$

- For 2.0×10^{-5} sec $< t \leq 10$ sec:

$$7.0 \times 10^{-4} \, k_1 k_2 \, t^{3/4}$$

$$= 7.0 \times 10^{-4} \, (1.25)\,(1) \, t^{3/4} \text{ J}$$

$$= 8.75 \times 10^{-4} \, t^{3/4} \text{ J}$$

- For 10 sec $< t \leq 198$ sec:

$$3.9 \times 10^{-3} \, k_1 k_2 \text{ J}$$

$$= 3.9 \times 10^{-3} \, (1.25)\,(1) \text{ J}$$

$$= 4.88 \times 10^{-3} \text{ J}$$

- For 198 sec $< t \leq 1.0 \times 10^4$ sec:

$$3.9 \times 10^{-3} \, k_1 k_2 \text{ J}$$

$$= 3.9 \times 10^{-3} \, (1.25)\,(0.00505\,t) \text{ J}$$

$$= 2.46 \times 10^{-5} \, t \text{ J}$$

- For $1 \times 10^4 \sec < t$:

$$3.9 \times 10^{-7} \, k_1 k_2 \, t \text{ J}$$

$$= 3.9 \times 10^{-7} \, (1.25)(50.5) \, t \text{ J}$$

$$= 2.46 \times 10^{-5} \, t \text{ J}$$

D) Accessible Emission Levels (Column 6 of Table C-1): Obtain accessible emission levels that correspond to the highest relative accessible emission levels.

- This is done by selecting the highest emission duration for the particular emission duration range:

$$2.0 \times 10^{-5} \sec$$

$$10 \sec$$

$$198 \sec$$

$$198 \sec < t < 1.0 \times 10^4 \sec$$

$$1.0 \times 10^4 \sec < t$$

Enter values in Column 8 of Table C-1 (Emission Duration—sec) From given information: Radiant Energy Level $= 1.30 \times 10^{-5} \, t$ J (i.e., Accessible Emission Level)

for $t = 2.0 \times 10^{-5}$ sec: $1.30 \times 10^{-5} (2.0 \times 10^{-5})$ J $= 2.6 \times 10^{-10}$ J

for $t = 10$ sec: $1.30 \times 10^{-5} (10)$ J $= 1.3 \times 10^{-4}$ J

for $t = 198$ sec: $1.30 \times 10^{-5} (198)$ J $= 2.57 \times 10^{-3}$ J

for $198 \sec < t < 1.0 \times 10^4$ sec: $1.30 \times 10^{-5} \, t$ J

for $1.0 \times 10^4 \sec < t$: $1.3 \times 10^{-5} \, t$ J

E) Relative Accessible Emission Levels $\left[\dfrac{E(\lambda, t)}{E(N, \lambda, t)_{\max}} \right]$ (Column 7 of Table C-1): Obtain ratios for respective values of Items (C) and (D) above.

$$\frac{2.6 \times 10^{-10} \text{ J}}{2.5 \times 10^{-7} \text{ J}} = 1.04 \times 10^{-3}$$

$$\frac{1.3 \times 10^{-4} \text{ J}}{8.75 \times 10^{-4} \, t^{3/4} \text{ J}} = \frac{1.3}{8.75 \, (10)^{3/4}} = \frac{1.49 \times 10^{-1}}{5.62} = 0.0264$$

$$\frac{2.57 \times 10^{-3} \text{ J}}{4.88 \times 10^{-3} \text{ J}} = 0.527$$

$$\frac{1.30 \times 10^{-5} \, t \text{ J}}{2.46 \times 10^{-5} \, t \text{ J}} = 0.528$$

$$\frac{1.30 \times 10^{-5} \, t \, J}{2.46 \times 10^{-5} \, t \, J} = 0.528$$

Note that all of the relative accessible emission levels are again below unity. Therefore, the 750-nm component of the laser radiation is below the accessible emission limits of the Class I level.

F) Estimated Error in Value Reported in Column 6 and Confidence Level (Column 11 of Table C-1)

$$\pm 15\% \text{ of } 2.60 \times 10^{-10} \, J: \pm 0.39 \times 10^{-10} \, J \, (3\sigma)$$

$$\pm 15\% \text{ of } 1.30 \times 10^{-4} \, J: \pm 0.195 \times 10^{-4} \, J \, (3\sigma)$$

$$\pm 15\% \text{ of } 2.57 \times 10^{-3} \, J: \pm 0.386 \times 10^{-3} \, J \, (3\sigma)$$

$$\pm 15\% \text{ of } 1.30 \times 10^{-5} \, t \, J: \pm 0.195 \times 10^{-5} \, t \, J \, (3\sigma)$$

$$\pm 15\% \text{ of } 1.30 \times 10^{-5} \, t \, J: \pm 0.195 \times 10^{-5} \, t \, J \, (3\sigma)$$

G) Relative Accessible Emission Levels for 15% Estimated Measurement Error

$$\frac{(2.6 \times 10^{-10} \, J) + (0.39 \times 10^{-10} \, J)}{2.5 \times 10^{-7} \, J} = 1.195 \times 10^{-3}$$

$$\frac{(1.3 \times 10^{-4} \, J) + (0.195 \times 10^{-4} \, J)}{8.75 \times 10^{-4} \, t^{3/4} \, J} = \frac{1.71 \times 10^{-1}}{(10)^{3/4}} = \frac{1.71 \times 10^{-1}}{5.62} = 0.0304$$

$$\frac{(2.57 \times 10^{-3} \, J) + (0.386 \times 10^{-3} \, J)}{4.88 \times 10^{-3} \, J} = 0.605$$

$$\frac{(1.30 \times 10^{-5} \, t \, J) + (0.195 \times 10^{-5} \, t \, J)}{2.46 \times 10^{-5} \, t \, J} = 0.607$$

$$\frac{(1.30 \times 10^{-5} \, t \, J) + (0.195 \times 10^{-5} \, t \, J)}{2.46 \times 10^{-5} \, t \, J} = 0.607$$

H) Miscellaneous Columns of Table C-1: Columns 1, 2, 9, 10, 12 and 13 are the same as for Step (2) above. Column 3 entry is 750 nm.

Step 4—Consider Wavelength Range Containing Only 1,100-nm Component

A) Wavelength-Dependent Factors k_1 and k_2 (From Table 7-2A)

$$k_1 = 5.0 \quad \text{for all values of } t$$

$$k_2 = 1.0 \quad \text{for } t \leq 100 \text{ sec}$$

$$\text{also } k_2 = 0.01t \quad \text{for } 100 \text{ sec} < t \leq 1.0 \times 10^4 \text{ sec}$$

$$\text{and } k_2 = 100 \quad \text{for } t > 10^4 \text{ sec}$$

B) Emission Duration Ranges (Column 4 of Table C-1): Examination of the accessible emission limits for the Class I level (Table 7-1A) and the above k_1 and k_2, indicates the emission duration ranges to be considered for $\lambda = 1,100$ nm are:

$$1.0 \times 10^{-9} \ \sec < t \le 2.0 \times 10^{-5}$$

$$2.0 \times 10^{-5} \ \sec < t \le 10 \ \sec$$

$$10 \ \sec < t \le 100 \ \sec$$

$$100 \ \sec < t \le 1.0 \times 10^4 \ \sec$$

$$1 \times 10^4 \ \sec < t$$

C) Accessible Emission Limits (Column 5 of Table C-1): From Table 7-1A, using appropriate k_1 and k_2 the accessible emission limits follow for the emission duration ranges shown:

- For $1.0 \times 10^{-9} \ \sec < t \le 2.0 \times 10^{-5} \ \sec$:

  $$2.0 \times 10^{-7} \ k_1 k_2 \ J$$

  $$= 2.0 \times 10^{-7} \ (5.0) \ (1.0) \ J$$

  $$= 1.0 \times 10^{-6} \ J$$

- For $2.0 \times 10^{-5} \ \sec < t \le 10 \ \sec$:

  $$7.0 \times 10^{-4} \ k_1 k_2 \ t^{3/4} \ J$$

  $$= 7.0 \times 10^{-4} \ (5.0) \ (1.0) \ t^{3/4} \ J$$

  $$= 3.5 \times 10^{-3} \ t^{3/4} \ J$$

- For $10 \ \sec < t \le 100 \ \sec$:

  $$3.9 \times 10^{-3} \ k_1 k_2 \ J$$

  $$= 3.9 \times 10^{-3} \ (5.0) \ (1.0) \ J$$

  $$= 1.95 \times 10^{-2} \ J$$

- For $100 \ \sec < t \le 1.0 \times 10^4 \ \sec$:

  $$3.9 \times 10^{-3} \ k_1 k_2 \ J$$

  $$= 3.9 \times 10^{-3} \ (5.0) \ (0.01 \, t) \ J$$

  $$= 1.95 \times 10^{-4} \ t \ J$$

- For $1 \times 10^4 \ \sec < t$:

  $$3.9 \times 10^{-7} \ k_1 k_2 \ t \ J$$

  $$= 3.9 \times 10^{-7} \ (5.0) \ (100) \ t \ J$$

  $$= 1.95 \times 10^{-4} \ t \ J$$

D) Accessible Emission Levels (Column 6 of Table C-1): Obtain accessible emission levels that correspond to the highest relative accessible emission levels.

- This is done by selecting the highest emission duration for the particular emission duration range:

$$2.0 \times 10^{-5} \text{ sec}$$

$$10 \text{ sec}$$

$$100 \text{ sec}$$

$$100 \text{ sec} < t < 1.0 \times 10^4 \text{ sec}$$

$$1.0 \times 10^4 \text{ sec} < t$$

Enter values in Column 8 of Table C-1 (Emission Duration–sec)
From given information: Radiant Energy Level = $1.60 \times 10^{-4} \ t$ J (i.e., Accessible Emission Level)

for $t = 2.0 \times 10^{-5}$ sec: $1.60 \times 10^{-4} (2.0 \times 10^{-5})$ J = 3.2×10^{-9} J

for $t = 10$ sec: $1.60 \times 10^{-4} (10)$ J = 1.6×10^{-3} J

for $t = 100$ sec: $1.60 \times 10^{-4} (100)$ J = 1.6×10^{-2} J

for 100 sec $< t < 1.0 \times 10^4$ sec: $1.6 \times 10^{-4} \ t$ J

for 1.0×10^4 sec $< t$: $1.6 \times 10^{-4} \ t$ J

E) Relative Accessible Emission Levels $\left[\dfrac{E(\lambda, t)}{E(N, \lambda, t)_{\max}} \right]$ (Column 7 of Table C-1): Obtain ratios for respective values of Items (C) and (D) above.

$$\frac{3.2 \times 10^{-9} \text{ J}}{1.0 \times 10^{-6} \text{ J}} = 3.2 \times 10^{-3} = 0.0032$$

$$\frac{1.6 \times 10^{-3} \text{ J}}{3.5 \times 10^{-3} \ t^{3/4} \text{ J}} = \frac{1.6}{3.5 \ (10)^{3/4}} = \frac{1.6}{(3.5) \ (5.62)} = 0.081$$

$$\frac{1.6 \times 10^{-2} \text{ J}}{1.95 \times 10^{-2} \text{ J}} = 0.820$$

$$\frac{1.6 \times 10^{-4} \ t \text{ J}}{1.95 \times 10^{-4} \ t \text{ J}} = 0.820$$

$$\frac{1.6 \times 10^{-4} \ t \text{ J}}{1.95 \times 10^{-4} \ t \text{ J}} = 0.820$$

F) Estimated Error in Value Reported in Column 6 and Confidence Level (Column 11 of Table C-1)

$\pm15\%$ of 3.2×10^{-9} J: $\pm0.48 \times 10^{-9}$ J (3σ)

$\pm15\%$ of 1.6×10^{-3} J: $\pm0.24 \times 10^{-3}$ J (3σ)

$\pm15\%$ of 1.6×10^{-2} J: $\pm0.24 \times 10^{-2}$ J (3σ)

$\pm15\%$ of 1.6×10^{-4} t J: $\pm0.24 \times 10^{-4}$ t J (3σ)

$\pm15\%$ of 1.6×10^{-4} t J: $\pm0.24 \times 10^{-4}$ t J (3σ)

G) Relative Accessible Emission Levels for 15% Estimated Measurement Error

$$\frac{(3.2 \times 10^{-9} \text{ J}) + (0.48 \times 10^{-9} \text{ J})}{1.0 \times 10^{-6} \text{ J}} = 0.00368$$

$$\frac{(1.6 \times 10^{-3} \text{ J}) + (0.24 \times 10^{-3} \text{ J})}{3.5 \times 10^{-3} \; t^{3/4} \text{ J}} = \frac{1.84}{(3.5)(5.65)} = 0.0935$$

$$\frac{(1.6 \times 10^{-2} \text{ J}) + (0.24 \times 10^{-2} \text{ J})}{1.95 \times 10^{-2} \text{ J}} = 0.944$$

$$\frac{(1.6 \times 10^{-4} \; t \text{ J}) + (0.24 \times 10^{-4} \; t \text{ J})}{1.95 \times 10^{-4} \; t \text{ J}} = 0.944$$

$$\frac{(1.6 \times 10^{-4} \; t \text{ J}) + (0.24 \times 10^{-4})}{1.95 \times 10^{-4} \; t \text{ J}} = 0.944$$

This shows that the 1,100-nm component is quite close to unity (0.944); however, even with the probable error considered, it is still Class I level.

H) Miscellaneous Columns of Table C-1: Columns 1, 2, 9, 10, 12 and 13 are the same as for Step (2) above. Column 3 entry is 1,100 nm.

Step 5—Consider Multiple Wavelengths (750 nm and 1,100 nm) in Same Range (400 nm to 1,400 nm)

• For $t > 198$ seconds, the maximum ratio of the accessible emission levels to the Class I limits is 0.528 for 750 nm, and the maximum ratio is 0.820 for the 1,100-nm component. Considering the 15% estimated measurement error the values are:

0.607 for 750-nm component

and 0.944 for 1,100-nm component

• Individually, the components are of Class I level, but the sums of the ratios are:
—Without measurement error: 0.528 + 0.820 = 1.345
—With measurement error: 0.607 + 0.944 = 1.551
Therefore, the composite radiation is greater than the Class I level.

- Since the sum of the ratios exceeds unity, we should test the accessible emission levels of the two components in the wavelength range of 400 nm $<$ $\lambda \leq$ 1,400 nm against Class IIIb limits to assure ourselves that the radiation is not of Class IV level. However, the highest relative accessible emission level for this wavelength range to the Class I limits is only 1.551. The Class IIIb limits are several orders of magnitude higher than the Class I limits. Thus, the calculations need not be carried through.
- In conclusion, from the parameters presented in Example 1, the radiation, because of the combined behavior of the components at 750 nm and 1,100 nm, is found to be of a Class IIIb level.

Index